Reviews of Environmental Contamination and Toxicology

VOLUME 205

For further volumes:
http://www.springer.com/series/398

Reviews of Environmental Contamination and Toxicology

Editor
David M. Whitacre

Editorial Board
María Fernanda Cavieres, Playa Ancha, Valparaíso, Chile • Charles P. Gerba, Tucson, Arizona, USA
John Giesy, Saskatoon, Saskatchewan, Canada • O. Hutzinger, Bayreuth, Germany
James B. Knaak, Getzville, New York, USA
James T. Stevens, Winston-Salem, North Carolina, USA
Ronald S. Tjeerdema, Davis, California, USA • Pim de Voogt, Amsterdam, The Netherlands
George W. Ware, Tucson, Arizona, USA

Founding Editor
Francis A. Gunther

VOLUME 205

Coordinating Board of Editors

Dr. David M. Whitacre, *Editor*
Reviews of Environmental Contamination and Toxicology

5115 Bunch Road
Summerfield North, Carolina 27358, USA
(336) 634-2131 (PHONE and FAX)
E-mail: dmwhitacre@triad.rr.com

Dr. Herbert N. Nigg, *Editor*
Bulletin of Environmental Contamination and Toxicology

University of Florida
700 Experiment Station Road
Lake Alfred, Florida 33850, USA
(863) 956-1151; FAX (941) 956-4631
E-mail: hnn@LAL.UFL.edu

Dr. Daniel R. Doerge, *Editor*
Archives of Environmental Contamination and Toxicology

7719 12th Street
Paron, Arkansas 72122, USA
(501) 821-1147; FAX (501) 821-1146
E-mail: AECT_editor@earthlink.net

ISSN 0179-5953
ISBN 978-1-4419-5622-4   e-ISBN 978-1-4419-5623-1
DOI 10.1007/978-1-4419-5623-1
Springer New York Dordrecht Heidelberg London

Library of Congress Control Number: 2009943238

© Springer Science+Business Media, LLC 2010
All rights reserved. This work may not be translated or copied in whole or in part without the written permission of the publisher (Springer Science+Business Media, LLC, 233 Spring Street, New York, NY 10013, USA), except for brief excerpts in connection with reviews or scholarly analysis. Use in connection with any form of information storage and retrieval, electronic adaptation, computer software, or by similar or dissimilar methodology now known or hereafter developed is forbidden.
The use in this publication of trade names, trademarks, service marks, and similar terms, even if they are not identified as such, is not to be taken as an expression of opinion as to whether or not they are subject to proprietary rights.

Printed on acid-free paper

Springer is part of Springer Science+Business Media (www.springer.com)

Foreword

International concern in scientific, industrial, and governmental communities over traces of xenobiotics in foods and in both abiotic and biotic environments has justified the present triumvirate of specialized publications in this field: comprehensive reviews, rapidly published research papers and progress reports, and archival documentations. These three international publications are integrated and scheduled to provide the coherency essential for nonduplicative and current progress in a field as dynamic and complex as environmental contamination and toxicology. This series is reserved exclusively for the diversified literature on "toxic" chemicals in our food, our feeds, our homes, recreational and working surroundings, our domestic animals, our wildlife, and ourselves. Tremendous efforts worldwide have been mobilized to evaluate the nature, presence, magnitude, fate, and toxicology of the chemicals loosed upon the Earth. Among the sequelae of this broad new emphasis is an undeniable need for an articulated set of authoritative publications, where one can find the latest important world literature produced by these emerging areas of science together with documentation of pertinent ancillary legislation.

Research directors and legislative or administrative advisers do not have the time to scan the escalating number of technical publications that may contain articles important to current responsibility. Rather, these individuals need the background provided by detailed reviews and the assurance that the latest information is made available to them, all with minimal literature searching. Similarly, the scientist assigned or attracted to a new problem is required to glean all literature pertinent to the task, to publish new developments or important new experimental details quickly, to inform others of findings that might alter their own efforts, and eventually to publish all his/her supporting data and conclusions for archival purposes.

In the fields of environmental contamination and toxicology, the sum of these concerns and responsibilities is decisively addressed by the uniform, encompassing, and timely publication format of the Springer triumvirate:

Reviews of Environmental Contamination and Toxicology [Vol. 1 through 97 (1962–1986) as Residue Reviews] for detailed review articles concerned with any aspects of chemical contaminants, including pesticides, in the total environment with toxicological considerations and consequences.

Bulletin of Environmental Contamination and Toxicology (Vol. 1 in 1966) for rapid publication of short reports of significant advances and discoveries in the fields of air, soil, water, and food contamination and pollution as well as methodology and other disciplines concerned with the introduction, presence, and effects of toxicants in the total environment.

Archives of Environmental Contamination and Toxicology (Vol. 1 in 1973) for important complete articles emphasizing and describing original experimental or theoretical research work pertaining to the scientific aspects of chemical contaminants in the environment.

Manuscripts for Reviews and the Archives are in identical formats and are peer reviewed by scientists in the field for adequacy and value; manuscripts for the *Bulletin* are also reviewed, but are published by photo-offset from camera-ready copy to provide the latest results with minimum delay. The individual editors of these three publications comprise the joint Coordinating Board of Editors with referral within the board of manuscripts submitted to one publication but deemed by major emphasis or length more suitable for one of the others.

Coordinating Board of Editors

Preface

The role of *Reviews* is to publish detailed scientific review articles on all aspects of environmental contamination and associated toxicological consequences. Such articles facilitate the often complex task of accessing and interpreting cogent scientific data within the confines of one or more closely related research fields.

In the nearly 50 years since *Reviews of Environmental Contamination and Toxicology* (formerly *Residue Reviews*) was first published, the number, scope, and complexity of environmental pollution incidents have grown unabated. During this entire period, the emphasis has been on publishing articles that address the presence and toxicity of environmental contaminants. New research is published each year on a myriad of environmental pollution issues facing people worldwide. This fact, and the routine discovery and reporting of new environmental contamination cases, creates an increasingly important function for *Reviews*.

The staggering volume of scientific literature demands remedy by which data can be synthesized and made available to readers in an abridged form. *Reviews* addresses this need and provides detailed reviews worldwide to key scientists and science or policy administrators, whether employed by government, universities, or the private sector.

There is a panoply of environmental issues and concerns on which many scientists have focused their research in past years. The scope of this list is quite broad, encompassing environmental events globally that affect marine and terrestrial ecosystems; biotic and abiotic environments; impacts on plants, humans, and wildlife; and pollutants, both chemical and radioactive; as well as the ravages of environmental disease in virtually all environmental media (soil, water, air). New or enhanced safety and environmental concerns have emerged in the last decade to be added to incidents covered by the media, studied by scientists, and addressed by governmental and private institutions. Among these are events so striking that they are creating a paradigm shift. Two in particular are at the center of ever-increasing media as well as scientific attention: bioterrorism and global warming. Unfortunately, these very worrisome issues are now superimposed on the already extensive list of ongoing environmental challenges.

The ultimate role of publishing scientific research is to enhance understanding of the environment in ways that allow the public to be better informed. The term "informed public" as used by Thomas Jefferson in the age of enlightenment conveyed the thought of soundness and good judgment. In the modern sense, being "well informed" has the narrower meaning of having access to sufficient information. Because the public still gets most of its information on science and technology from TV news and reports, the role for scientists as interpreters and brokers of scientific information to the public will grow rather than diminish. Environmentalism is the newest global political force, resulting in the emergence of multinational consortia to control pollution and the evolution of the environmental ethic. Will the new politics of the 21st century involve a consortium of technologists and environmentalists, or a progressive confrontation? These matters are of genuine concern to governmental agencies and legislative bodies around the world.

For those who make the decisions about how our planet is managed, there is an ongoing need for continual surveillance and intelligent controls to avoid endangering the environment, public health, and wildlife. Ensuring safety-in-use of the many chemicals involved in our highly industrialized culture is a dynamic challenge, for the old, established materials are continually being displaced by newly developed molecules more acceptable to federal and state regulatory agencies, public health officials, and environmentalists.

Reviews publishes synoptic articles designed to treat the presence, fate, and, if possible, the safety of xenobiotics in any segment of the environment. These reviews can be either general or specific, but properly lie in the domains of analytical chemistry and its methodology, biochemistry, human and animal medicine, legislation, pharmacology, physiology, toxicology, and regulation. Certain affairs in food technology concerned specifically with pesticide and other food-additive problems may also be appropriate.

Because manuscripts are published in the order in which they are received in final form, it may seem that some important aspects have been neglected at times. However, these apparent omissions are recognized, and pertinent manuscripts are likely in preparation or planned. The field is so very large and the interests in it are so varied that the editor and the editorial board earnestly solicit authors and suggestions of underrepresented topics to make this international book series yet more useful and worthwhile.

Justification for the preparation of any review for this book series is that it deals with some aspect of the many real problems arising from the presence of foreign chemicals in our surroundings. Thus, manuscripts may encompass case studies from any country. Food additives, including pesticides, or their metabolites that may persist into human food and animal feeds are within this scope. Additionally, chemical contamination in any manner of air, water, soil, or plant or animal life is within these objectives and their purview.

Manuscripts are often contributed by invitation. However, nominations for new topics or topics in areas that are rapidly advancing are welcome. Preliminary communication with the editor is recommended before volunteered review manuscripts are submitted.

Summerfield, NC, USA   David M. Whitacre

Contents

Gammarus **spp. in Aquatic Ecotoxicology and Water Quality Assessment: Toward Integrated Multilevel Tests** 1
Petra Y. Kunz, Cornelia Kienle, and Almut Gerhardt

The Svalbard Glaucous Gull as Bioindicator Species in the European Arctic: Insight from 35 Years of Contaminants Research 77
J. Verreault, G.W. Gabrielsen, and J.O. Bustnes

Fenamiphos and Related Organophosphorus Pesticides: Environmental Fate and Toxicology 117
Tanya Cáceres, Mallavarapu Megharaj, Kadiyala Venkateswarlu, Nambrattil Sethunathan, and Ravi Naidu

Index 163

Contributors

J.O. Bustnes Norwegian Institute for Nature Research, Polar Environmental Centre, NO-9296 Tromsø, Norway

Tanya Cáceres Centre for Environmental Risk Assessment and Remediation (CERAR), Cooperative Research Centre for Contamination Assessment and Remediation of the Environment (CRC CARE), University of South Australia, Mawson Lakes 5095, SA, Australia

G.W. Gabrielsen Norwegian Polar Institute, Polar Environmental Centre, NO-9296 Tromsø, Norway

Almut Gerhardt Ecotox Centre, Swiss Center for Applied Ecotoxicology, Eawag/EPFL, Überlandstrasse 133, CH-8600 Dübendorf, Switzerland

Cornelia Kienle Ecotox Centre, Swiss Center for Applied Ecotoxicology, Eawag/EPFL, Überlandstrasse 133, CH-8600 Dübendorf, Switzerland

Petra Y. Kunz Ecotox Centre, Swiss Center for Applied Ecotoxicology, Eawag/EPFL, Überlandstrasse 133, CH-8600 Dübendorf, Switzerland

Mallavarapu Megharaj Centre for Environmental Risk Assessment and Remediation (CERAR), Cooperative Research Centre for Contamination Assessment and Remediation of the Environment (CRC CARE), University of South Australia, Mawson Lakes Boulevard, Mawson Lakes 5095, SA, Australia

Ravi Naidu Centre for Environmental Risk Assessment and Remediation (CERAR), Cooperative Research Centre for Contamination Assessment and Remediation of the Environment (CRC CARE), University of South Australia, Mawson Lakes Boulevard, Mawson Lakes 5095, SA, Australia

Nambrattil Sethunathan Flat No. 103, Ushodaya Apartments, Sri Venkateswara Officers Colony, Ramakrishnapuram, Secunderabad 500056, India

Kadiyala Venkateswarlu Sri Krishnadevaraya University, Anantapur 515055, India

J. Verreault National Wildlife Research Centre, Carleton University, Ottawa, Ontario, K1A 0H3, Canada

Gammarus spp. in Aquatic Ecotoxicology and Water Quality Assessment: Toward Integrated Multilevel Tests

Petra Y. Kunz, Cornelia Kienle, and Almut Gerhardt

Contents

1 Introduction 2
2 Culturing of Gammarids 5
3 Gammarids in Lethality Testing 7
   3.1 Pesticides, Metals, and Surfactants 7
   3.2 Extracted, Fractionated Sediments 8
   3.3 Coastal Sediment Toxicity 8
4 Feeding Activity 10
   4.1 Time–Response Feeding Assays 10
   4.2 Food Choice Experiments 16
   4.3 Leaf-Mass Feeding Assays Linked to Food Consumption 17
   4.4 Modeling of Feeding Activity and Rate 19
   4.5 Post-exposure Feeding Depression Assay 20
   4.6 Effects of Parasites on Gammarid Feeding Ecology 21
5 Behavior 22
   5.1 Antipredator Behavior 22
   5.2 Multispecies Freshwater Biomonitor® (MFB) 27
   5.3 A Sublethal Pollution Bioassay with Pleopod Beat Frequency and Swimming Endurance 29
   5.4 Behavior in Combination with Other Endpoints 29
6 Mode-of-Action Studies and Biomarkers 31
   6.1 Bioenergetic Responses, Excretion Rate and Respiration Rate 32
   6.2 Population Experiments, and Development and Reproduction Modeling . . . 41
   6.3 Endpoints and Biomarkers for Endocrine Disruption in Gammarids 44
   6.4 Other Specific Biomarkers for Detecting Multiple Stressors in Gammarids . . . 48
7 Exposure Types 53
   7.1 Pulsed Exposure Assays and Models 53

P.Y. Kunz (✉)
Ecotox Centre, Swiss Center for Applied Ecotoxicology, Eawag/EPFL, Überlandstrasse 133, CH-8600 Dübendorf, Switzerland
e-mail: petra.kunz@oekotoxzentrum.ch

D.M. Whitacre (ed.), *Reviews of Environmental Contamination and Toxicology*,
Reviews of Environmental Contamination and Toxicology 205,
DOI 10.1007/978-1-4419-5623-1_1, © Springer Science+Business Media, LLC 2010

7.2 Sediment Toxicity Assays 57
7.3 In Situ Tests 58
8 Discussion 62
 8.1 Evaluation of Existing Methods 62
 8.2 Perspectives on a Multimetric Gammarus spp. Test System 64
9 Summary 65
References 66

1 Introduction

More than 4500 species belong to the crustacean sub-order Gammaridea (order Amphipoda) (Bousfield 1973). Among Amphipods, the Gammaridea are the most widespread group and are found throughout a range of marine, freshwater, and terrestrial habitats (Bousfield 1973; Lincoln 1979), whereas the three other amphipod sub-orders (Hyperiidea, Ingolfiellidea, and Caprellidea) are highly specialized and ecologically restricted. *Gammarus* is the amphipod genus with the highest number of epigean freshwater species, comprising over 100 species that are distributed throughout the Northern Hemisphere (Karaman and Pinkster 1977). Abiotic factors such as temperature, salinity, oxygen, acidity, and pollution play an important role in the distribution of *Gammarus* species (Whitehurst and Lindsey 1990) and members of this species are often found in great abundance under rocks, in gravel, or in coarse substrates and among living and dead vegetation (Fitter and Manuel 1994). These substrata provide both shelter from predators and a supply of organic detritus and other foodstuffs, with the result that in many riverine communities, amphipod species such as *Gammarus pulex* (Linnaeus) may represent the dominant macroinvertebrate in terms of biomass (Macneil et al. 1997; Shaw 1979).

The species *G. fossarum* and *G. pulex,* for example, are widespread and functionally important in streams throughout much of Europe and Northern Asia (Karaman and Pinkster 1977). They display a wide trophic repertoire, feeding as herbivores, detritivores, and predators. Stream-conditioned leaves, biofilms that grow on them, dead chironomids, live juvenile isopods, and even juvenile and wounded/trapped fish are part of their diet (Fielding et al. 2003; Macneil et al. 1997). Intraguild predation and cannibalism have been observed in many amphipod species, and data from *G. pulex* suggest that this is more common than previously realized (Macneil et al. 1997). This "foraging plasticity" is linked to the success of *Gammarus* spp. as they persist in colonizing and invading disturbance-prone ecosystems (Macneil et al. 1997). Hence, the ecological value of a macroinvertebrate like *G. pulex* in streams may exceed its important role as a shredder in processing leaf material that falls into streams (Welton et al. 1983; Willoughby and Earnshaw 1982; Willoughby and Sutcliffe 1976), and as prey items for many fish species (Maitland 1966; Smyly 1957; Welton 1979).

Gammarus species have a complex life cycle, which is of value in ecotoxicological studies because changes in mating behavior can be observed more easily in the presence of xenoestrogen exposure (Segner et al. 2003; Watts et al. 2001, 2002, 2003). *Gammarus* females are available for mating only during a brief period directly after the molt. Alternatively, males are available for mating during most of the molt cycle (Sutcliffe 1993), which results in a male-biased operational sex ratio. To address this situation, males engage in precopulatory mate guarding when encountering a female nearing the molt stage (Ridley 1983). Prior to mating, males grab and hold different females before deciding which one is likely to produce the most eggs. Pairing in *Gammarus* is positively size assortive and, unlike other arthropods, males are larger than females (Dick and Elwood 1996; Thomas et al. 1995; Zielinski 1998). When the male has found a suitable female, they form a precopula pair. The male holds the female under and parallel to his body using the first pair of gnathopods (Borowski 1984), and while carrying her, he performs all the necessary swimming movements (Bollache and Cezilly 2004). Pairs can remain in precopula for up to 2 weeks (Hartnoll and Smith 1980). As soon as the female sheds her skin, the male can mate with her. Then, the precopula pair separates and the female carries the developing eggs in her brooding pouch. The young hatch after 1–3 weeks; thereafter, the juveniles remain in the brooding pouch until the next female molt. Then, after 4–6 weeks, the young *Gammarus* swim out of the brooding pouch. Freshly hatched juveniles feed by coprophagy (feces of adults) (McCahon and Pascoe 1988b). The diet thereafter expands to include conditioned leaves, and after 1 month or so, the young feed only on conditioned leaves. The juveniles will, themselves, mate 3–4 months later, after they have reached sexual maturity and completed about 10 molts (McCahon and Pascoe 1988a). Gammarids can reach ages of 1–2 years.

The presence of gammarids in freshwater streams is crucial, because macroinvertebrate feeding is a major rate-limiting step in the processing of stream detritus (Cummins and Klug 1979). Detritus is an integrated decomposition of leaf material in streams and is brought about by a combination of chemical leaching, microbial decomposition (primarily by aquatic hyphomycetes), macroinvertebrate feeding, and physical abrasion (Webster and Benfield 1986). Environmental contaminants can reduce detritus processing by decreasing the microbial conditioning and/or the abundance or the feeding activity of detritivores such as *G. pulex* (Forrow and Maltby 2000; Webster and Benfield 1986). Reductions in gammarid abundance from increased mortality have, for example, been observed after exposing them to acutely toxic landfill leachates that contain environmental contaminants (Bloor et al. 2005). Toxicant-induced reductions in feeding rate can result in reduced growth, size, fecundity, and survival of individuals (Anderson and Cummins 1979; Maltby and Naylor 1990), thereby affecting the stream community structure (Sutcliffe and Hildrew 1989).

Feeding activity is one of many behavioral responses that may be affected by environmental contaminants. For example, changes in locomotory or ventilation behavior are compensatory, reversible adaptive responses to pollutants that may mitigate potential overt effects (e.g., direct behavioral response after perception of stress). Irreversible effects of a toxicant on a behavioral mechanism or expression are also observed in the behavioral response of an organism, after the toxicokinetic and toxicodynamic processes have started [e.g., acetylcholinesterase (AChE) inhibition exerted by neurotoxins; Gerhardt 1995].

Using behavioral parameters in ecotoxicology studies have advantages, to wit short response times (i.e., early warning responses), sensitivity (i.e., for neuromuscular toxins), non-invasiveness, ecological relevance, and the possibility for time-dependent data analysis (Fossi 1998; Gerhardt 2007; Scherer 1992). Changes in behavior may be used as important indicators for ecosystem health, because they rest on biochemical processes, but also reflect the fitness of the individual organism as well as potential effects on the population level, such as altered abundance of the species in the ecosystem. Behavioral responses seem to be of similar sensitivity and efficiency as biochemical and physiological responses and because of their indestructibility, continuous long-term monitoring is possible (Gerhardt 2007; Scherer 1992). To study changes in behavior that result from contaminant exposure is therefore an essential part of behavioral science, which can be called behavioral ecotoxicology (Gerhardt 2007).

In aquatic ecotoxicology, behavioral endpoints have been applied for fish, crayfish, copepods, and gammarids for about 20 years (Atchison et al. 1987; Beitinger 1990). In the case of gammarids, behavioral alteration tests allow one to measure several endpoints related to population structure, population density, and to inter- and intra-specific interactions. Such endpoints may be sensitive indicators of chemical stressors when used for biomonitoring purposes, for instance, biomonitoring with impedance conversion (Gerhardt 1995; Gerhardt et al. 1998), which can be used to quantitatively record behaviors such as ventilation, grazing, filter feeding, net spinning, and locomotion. It is known that *Gammarus* species can locate their food (De Lange et al. 2005) and detect predators through chemical cues from fish and injured conspecifics. Changes in behavior such as hiding in response to predators (Åbjörnsson et al. 2004; Baumgärtner et al. 2002; Gerhardt and Quindt 2000; Williams and Moore 1985; Wisenden et al. 1999; Wisenden et al. 2001; Wudkevich et al. 1997) or avoidance of chemical stress from exposure to pollution pulses (Gerhardt 1995, 1996; Gerhardt et al. 1998; Gerhardt and Quindt 2000) are crucial to optimize the chance of survival. Reproductive behaviors can also be examined; an example is the ability of males and females to detect each other, form precopulatory guarding pairs during the premating period, and display guarding behavior (Watts et al. 2001).

In addition to the vast published literature that addresses the effect of pollutants on *Gammarid* mortality, feeding, and behavior, many publications also address the effects of environmental pollutants on population structure, the endocrine system, stress response, the neural system, and bioenergetics, etc. Of those, only a few address the endocrine disruption in *Gammarids*; key endpoints investigated in these endocrine disruption studies include structure, size, length–frequency distributions, adult sex ratio, number of precopula pairs/ovigerous females, and secondary sex characteristics (Watts et al. 2002). Recently, different biomarkers, e.g., vitellogenin (Vtg) or heat shock proteins (hsp), have been used to investigate endocrine disruption in gammarids (De Coen and Janssen 2003; Gagné et al. 2005; Schirling et al. 2004). Moreover, biomarkers have increasingly been used to assess stress response, oxidative stress, exoskeleton integrity, and neurotoxicity (Correia et al. 2002; Scheil et al. 2008; Xuereb et al. 2007).

The above-mentioned toxicity endpoints have been utilized to address effects at the individual and population level; however, there have also been investigations on the structure and composition of communities that make important contributions. In this context, neozoa may play an important role in disturbed or polluted aquatic ecosystems. For example, *Dikerogammarus villosus*, a Ponto-Caspian species, is known to be a particularly successful invader and is currently the prevailing invasive gammarid species in large bodies of water in Southern Germany (Kinzler et al. 2008). Field observations suggest that *D. villosus* has replaced the native *G. pulex* and the invasive *D. haemobaphes* in some spans of the German Danube. Such shifts in species composition may help explain the long-term effects of pollutants in those areas and the species sensitivity differences observed for these pollutants.

The widely investigated *gammarids*, in particular species such as *G. pulex* and *G. fossarum*, are important members of freshwater ecosystems. Because they are sensitive to pollutants and other disturbances, *gammarids* may be valuable indicators for ecosystem health in aquatic freshwater ecotoxicology, especially because they have been utilized in manifold exposure scenarios (e.g., in situ, ex situ, via sediment, and to pulsed exposures).

The purpose of this review is to collect available data, methods, and biomarkers cogent to *Gammarus* spp. and investigate the potential of gammarids to serve as an emerging test species for freshwater ecosystems. Gammarids may fill a crucial gap in the assessment of aquatic ecosystem health, which is not yet filled by OECD-proposed test species, because *Gammarus* spp. naturally occur in streams of the Northern Hemisphere. The breadth of ecotoxicological studies published on gammarids suggests that they may be suitable test organisms for a more integrative ecotoxicity testing in situ and ex situ, and consequently the data gained from their use may be more ecologically relevant. In this review, we aim to provide an overview on the status of ecotoxicological testing with gammarids and to suggest avenues for continuing, combining, and integrating future gammarid research in aquatic ecotoxicology.

2 Culturing of Gammarids

To propagate gammarids for broader use as test organisms in aquatic ecotoxicology, the ability to culture them is an essential prerequisite. Few publications exist on how to culture gammarids. Most studies that utilize gammarids address adverse effects of environmental contaminants on them and obtain test specimens by collecting wild animals from uncontaminated, clean sites. Only a limited number of studies have used lab-cultured animals (Bloor et al. 2005; McCahon and Pascoe 1988a), indicating that the culturing and reproduction of gammarids in the lab might be rather difficult. To our knowledge, only three publications exist that describe in detail how to culture *G. pulex* (McCahon and Pascoe 1988a, 1988b; Bloor 2009). These publications explain that a minimum of 100 precopula pairs and 200 visibly gravid *G. pulex* females from an unpolluted source are needed for culturing purposes. For culturing, *G. pulex* are placed in 1-L plastic containers, each with a nylon mesh base through which juveniles can pass after being released from the brood pouch. These containers are suspended in an 8-L rearing tank with a flow-through supply of dechlorinated, aerated tap water. The rearing tank may contain water held under static conditions, if the water is periodically renewed. A 12-hour photoperiod, with a light intensity of 750 lux at the water surface, is used. Adults are fed on conditioned, common leaves (Sutcliffe et al. 1981), which are collected during the autumn, are air-dried, and stored until shortly before use. To incorporate bacteria and fungi into the gammarid diet, leaf material is then conditioned for at least 10 days in organically enriched dechlorinated water to initiate microbial breakdown (Kaushik and Hynes 1971; Willoughby and Sutcliffe 1976). Adults may also be fed on algae (Moore 1975). After 5 days, the breeding container is removed and several newly hatched individuals (500–1000) will be present in the rearing tank. Conditioned leaves are added to the rearing tanks for cover and, eventually, as food. Early hatched juveniles feed upon adult feces, which require that feces be supplied (by pipetting), after removal of the breeding containers; thereafter, after approximately 25 days, the animals can feed entirely on leaves.

McCahon and Pascoe (1988b) monitored growth by removing 40 gammarids from each growing tank at regular intervals, measuring the total body length microscopically (anterior margin of the head to posterior margin of the telson) to the nearest 0.25 mm, and by counting the number of segments on the primary flagella of each antenna. Juveniles that are released from the brood pouch possess five segments on the primary flagellus of each antenna. The number segments increase as growth progresses and a growth curve can be plotted, which allows calculation of the age of the cultured specimens. Approximately 70% of cultured juveniles survive to reach sexual maturity in time periods that are dependent on rearing temperature as follows: 130 days at 13°C (14–16 antennae segments, after 10 molts; McCahon and Pascoe 1988b), 120 days at 15–20°C (Hynes 1955), and 133 days at 15°C (Welton and Clarke 1980). Sexes can then be distinguished by visible identification of genital papillae in males and by fully developed oostegites in females; such oostegites have long fringed bristles, which interlace with one another to form the brood pouch. Females can produce 2–5 broods with a mean of 16 eggs each. By increasing rearing temperature and providing excess food, it is possible to culture animals throughout the year and attain a reduced time to sexual maturity.

For toxicity studies, McCahon and Pascoe (1988b) suggest using a test population that comprises organisms of mixed stages and ages. Therefore, at periodic intervals, the breeding adults in each container are transferred to fresh rearing tanks, and the newly hatched juveniles are gently removed from females by prodding the brood pouch. The authors suggest using the following age groups (in days) for toxicity testing: 4.5–9.5, 22.5–27.5, 44.5–49.5, 64.5–69.5, 79.5–84.5, and 217.5–222.5. McCahon and Pascoe (1988b) propose using 20 cultured *G. pulex* of each age class as well as 20 field-collected animals of unknown age in toxicity tests.

3 Gammarids in Lethality Testing

Gammarids (e.g., *G. pulex*) are widely used in experimental toxicity tests (McCahon and Pascoe 1988b) because of their well-known sensitivity to a wide range of pollutants and the fact that they are among the most sensitive aquatic invertebrates (Bloor et al. 2005; Cold and Forbes 2004; Mian and Mulla 1992; Van Wijngaarden et al. 2004; Wogram and Liess 2001).

3.1 Pesticides, Metals, and Surfactants

Multiple published studies exist on the acute toxicity of a wide range of chemicals and natural water samples toward gammarids. McCahon and Pascoe (1988b) exposed *G. pulex* to a range of cadmium chloride solutions (0, 10, 30, 50, 100, 300, and 1000 μg/L $CdCl_2 \bullet 2^1/_2H_2O$) and found that the median lethal time to mortality (LT_{50}) for all age classes decreased with increasing cadmium concentration. The slopes of the mortality time curves differed little among age groups or cadmium concentrations. The associated median lethal concentration (LC_{50}) values were very similar for most age classes, except for the oldest (220 days, 48 hour LC_{50} 4.7 mg Cd/L) and youngest (1 day, 48 hour LC_{50} 0.019 mg Cd/L) animals; juveniles were found to be almost 250 times more sensitive than those in the oldest age class. Pantani et al. (1997) investigated the acute toxicity of some common pesticides, metals, and surfactants to the amphipods *G. italicus* and *Echinogammarus tibaldii*. LC_{50} values of 16 insecticides varied from less than 1 μg/L for azinphos-methyl to several milligrams per liter for dimethoate. The sensitivity of *G. italicus* and *E. tibaldii* toward three herbicides and three surfactants was found to be about the same order of magnitude, and not very high (Pantani et al. 1997). Their toxicity findings for atrazine (LC_{50}: *G. italicus* 10.1 mg/L, *E. tibaldii* 3.3 mg/L) are very similar to previously observed acute toxicities in *G. pulex* and in *G. fasciatus* (Macek et al. 1976; Taylor et al. 1991). Metals showed differences in toxicity rankings between *G. italicus* (Zn<Cr<Cd<Hg<Cu) and *E. tibaldii* (Zn<Cd<Cu<Hg<Cr). Comparing these findings to other acute toxicity data for metals, mainly cadmium and copper, *G. fossarum* and *G. pulex* appear to be the most sensitive to cadmium (Musko et al. 1990; Williams and Moore 1985). In another study, several freshwater insects and crustacean species were exposed for 24 hour to the neonicotinoid insecticide thiacloprid. Among the investigated species, an increase in sensitivity, distributed over three orders of magnitude, was found: *Daphnia magna* < *Asellus aquaticus* = *G. pulex* < *Sympetrum striolatum* < *Culex pipiens* = *Notidobia ciliaris* = *Simulium latigonium*, with median lethal concentrations (LC_{50}s) of 4, 400, 153, 190, 31.2, 6.78, 5.47, and 5.76 µg/L, respectively (post-exposure observation 11–30 days; Beketov and Liess 2008).

3.2 Extracted, Fractionated Sediments

Boxhall and Maltby (1995) assessed the toxicity of sediment contaminated with road runoff by exposing *G. pulex* for 14 days to different extracted fractions of the contaminated sediment. The sediment was extracted with dichloromethane and then fractioned into three portions with increasing polarity using alumina–silica column chromatography. The fractions were analyzed using GC–MS (gas chromatography–mass spectrometry), GC/MS/IR (infrared), and IR spectrophotometry. Results revealed that the first fraction contained aliphatic hydrocarbons, the second fraction 2–5-ring polycyclic aromatic hydrocarbons (PAHs), and the third fraction (FC) substituted phenols and 4- and 5-ring PAHs. The 2–5-ring PAH fraction was toxic to *G. pulex* (mortality > 80%), whereas the other two fractions did not produce signs of toxicity after 14 days of exposure (Boxall and Maltby 1995).

3.3 Coastal Sediment Toxicity

Costa et al. (1998) used the marine amphipod *G. locusta* to assess coastal sediment toxicity in a 10-day static toxicity (mortality) test performed with laboratory-produced juveniles at 15°C and at a salinity of 33–34. *G. locusta* is a common species along European coastal areas and can be raised easily under laboratory conditions. This species is tolerant to a broad range of sediment types and toxins, including heavily contaminated field sediments (Costa et al. 1998), copper (LC_{50} = 56.8 mg Cu/kg dry weight, 0.9% total volatile solids), and to the gamma isomer of hexachlorocyclohexane (lindane) in the sediment (LC_{50} = 560.5 µg lindane/kg dry weight, 2% total volatile solids). The overall assay performance was identical to the American Society for Testing and Materials (ASTM) standard for sediment toxicity tests observed for marine and estuarine amphipods (ASTM 1993), revealing the utility of *G. locusta* in such tests.

The above-mentioned acute toxicity studies, as well as those shown in Table 1, indicate that gammarids are among the most sensitive organisms toward metal pollution. They show increased toxicity toward the neonicotinoid insecticide thiacloprid and the 2–5-ring PAHs but are not affected by PAHs with 4–5 rings. In some cases, 1-day-old juveniles were as much as 250 times more sensitive toward acute toxic effects than were 220-day-old adults and therefore should be increasingly used for such assessments. Compared to daphnids, juvenile and adult gammarids were more sensitive to some pesticides (e.g., lindane, thiacloprid) and copper (Table 1).

Table 1 Median lethal concentrations (LC_{50}s) of test chemicals for freshwater invertebrates and fish. For comparability reasons regarding sensitivity of the different species, the LC_{50} of the most sensitive species is indicated (gammarids: bold, daphnids: underlined)

Toxicant	Organism	Life stage	Time (hour)	LC_{50} (mg/L)	Source
Cadmium chloride	G. pulex	Juveniles	48	**0.019**	McCahon and Pascoe (1988a)
	G. pulex	Adults	48	**4.700**	McCahon and Pascoe (1988a)
	G. italicus	Adults	96	**9.1**	Pantani et al. (1997)
	E. tibaldii	Adults	96	**1.1**	Pantani et al. (1997)
	G. pulex	7–9 mm long	96	0.0821	Felten et al. (2008)
	G. pulex	7–9 mm long	120	0.0371	Felten et al. (2008)
	G pulex	7–9 mm long	168	0.0216	Felten et al. (2008)
	G. pulex	7–9 mm long	264	0.0105	Felten et al. (2008)
3,4-Dichloroaniline	D. magna	Larva, 1 mm	96	<u>0.16</u>	Adema and Vink (1981)
	D. magna	Adult	96	<u>1.0</u>	Adema and Vink (1981)
	Salmo gairdneri	Juvenile	96	2.7	Crossland (1988)
	Pimephales promelas	28–34 days	96	7.0–8.1	Call et al. (1987)
	G. pulex	2–3 molts	240	**5.0**	Taylor et al. (1991)
	Chironomus riparius	2nd instar	96	7.4	Taylor et al. (1991)
	Poecilia reticulata	Young	96	8.7	Adema and Vink (1981)
	G. pulex	2–3 molt	48	**17.4**	Taylor et al. (1991)
	Dreissena polymorpha	Adult, 1 cm	48	22	Adema and Vink (1981)
Atrazine	C. tentans	1st instar	48	0.72	Macek et al. (1976)
	G. fasciatus	1st molt	48	**5.7**	Macek et al. (1976)
	Salvelinus fontinalis	–	96	6.3	Macek et al. (1976)
	D. magna	<24-hour old	48	<u>6.9</u>	Macek et al. (1976)
	G. pulex	2–3 molts	96	**14.9**	Taylor et al. (1991)
	P. promelas	–	96	15.0	Macek et al. (1976)
	C. riparius	2nd instar	96	>30	Taylor et al. (1991)
Zinc chloride	G. italicus	Adults	96	**8.8**	Pantani et al. (1997)
	E. tibaldii	Adults	96	**25.9**	Pantani et al. (1997)
Copper	G. pseudolimnaeus	Adult	96	**0.020**	Arthur and Leonard (1970)
	G. pulex	2–3 molts	96	**0.037**	Girling et al. (2000)
	D. magna	<24-hour old	72	<u>0.09</u>	Winner and Farrell (1976)
	S. gairdneri	–	72	0.4	Brown (1968)
	C. riparius	2nd instar	96	0.7	Taylor et al. (1991)
	C. decorus	4th instar	48	0.74	Kosalwat and Knight (1987)
	P. promelas	–	96	2.2	Brungs et al. (1976)

Table 1 (continued)

Toxicant	Organism	Life stage	Time (hour)	LC$_{50}$ (mg/L)	Source
Lindane	C. riparius	2nd instar	96	0.034	Taylor et al. (1991)
	G. pulex	Adult	96	**0.034**	Abel (1980)
	G. fasciatus	3rd molt	48	**0.039**	Macek et al. (1976)
	S. fontinalis	–	48	0.044	Macek et al. (1976)
	B. rhodani	Larva	96	0.054	Green et al. (1986)
	G. pulex	2–3 molt	96	**0.079**	Taylor et al. (1991)
	P. promelas	–	48	>0.100	Macek et al. (1976)
	C. tentans	1st instar	48	0.207	Macek et al. (1976)
	G. pulex	Adult	96	**0.225**	Green et al. (1986)
	C. riparius	4th instar	96	0.235	Green et al. (1986)
	D. magna	<24-hour old	48	0.485	Macek et al. (1976)
Bisphenol A	G. pulex	Adult	240	1.49	Watts et al. (2001)
Ethinylestradiol	G. pulex	Adult	240	0.84	Watts et al. (2001)
Esfenvalerate	G. pulex	Adult	48	0.00014	Cold and Forbes (2004)
	G. pulex	Juvenile	48	0.00013	Cold and Forbes (2004)
	D. magna	Juvenile	48	0.00027	Fairchild et al. (1992)
Fenoxycarb	G. fossarum	Adult	96	10	Schmidt (2003)
	G. fossarum	Juveniles	96	4	Schmidt (2003)

4 Feeding Activity

One approach to assess water quality is to use feeding activity/inhibition of macroinvertebrates as a measure for a wide range of stressors. Such tests can be measured ex situ and in situ and have been proposed as good indicators for water quality (Crane and Maltby 1991). Table 2 provides an overview of different feeding activity test types employed with gammarids. Feeding rate is an advantageous endpoint because it is based on sublethal responses of a single species and is therefore more sensitive. Moreover, the use of this endpoint gives a more rapid response, than does community-based measures that require species eradication before an impact is detected (Maltby et al. 2002). However, feeding rate is influenced by intrinsic factors, i.e., parasite load (Pascoe et al. 1995), source population (Crane et al. 1995; Maltby and Crane 1994; Veerasingham and Crane 1992), and body size (Nilsson 1974), and extrinsic factors, i.e., temperature, dissolved oxygen content (Maltby et al. 1990), and pH (Naylor et al. 1989). Because feeding rate is variable, depending on the status of the organism and its surrounding environment, understanding those intrinsic factors and causes for variability is crucial for water quality assessment.

4.1 Time–Response Feeding Assays

Taylor et al. (1993) proposed a feeding assay designed to quantify the feeding activity of *G. pulex* by utilizing a time–response analysis of the feeding of amphipods on eggs of *Artemia salina*. In this assay, individual test organisms are transferred

Table 2 Test methods that utilize gammarids to assess toxicant effects on feeding activity

Species	Test substance/media	Type of exposure	Exposure period	Endpoints/biomarkers	Effect/effect concentration	Remarks	Reference
G. pulex males > 9 mm	Copper (Cu)	Feeding activity, aqueous static short-term exposures	3, 20, 48, and 96 hour	Median feeding time (MFT) on *A. salina* eggs	Significant increase in MFD: 3 hour: >101 µg/L Cu; 20 hour: >72 µg/L Cu; 48 hour: >21.5 µg/L Cu; 96 hour: >8.3 µg/L Cu		Taylor et al. (1993)
G. pulex juveniles, 5 mm	Cu; Lindane (LD), 3,4-dichloroaniline (DCA)	Feeding activity, static short-term exposures	96, 240 hour	Feeding activity (FT_{50}) on *A. salina* eggs	Significant reduced FT_{50} after 96 hour: >12.1 µg/L Cu; >8.4 µg/L LD; >918 µg/L DCA	Significantly increased FT_{50} after 240 hour exposure to 0.09 µg/L LD – stimulatory effects at low concentrations	Blockwell et al. (1998)
G. fossarum	Antibiotic mixture: sulfamethoxazole, trimethoprim, erythromycin-H_2O, roxithromycin, clarithromycin	Static food choice experiment with conditioned leaf discs, +/- antibiotics (total conc. 2, 200 µg/L);	48 hour	Food choice	Food choice: 200 µg/L ant bacterial conditioned leaves significantly preferred over control leafs; same tendency for 2 µg/L antibiotic leaves	Number of bacteria on antibiotic leaves did not differ from controls, but fungal biomass was significantly higher at 200 µg/L conditioned discs	Bundschuh et al. (2009)
G. fossarum adult males	Highly acidic stream vs. reference stream, with conditioned leaves	In situ: consummation of leaf mass	6 days	Consummation of leaf mass: dry weight, leaf-disk area	Feeding activity: 113 and 46 times lower on exposed and control leaves in acidic vs. control stream	Almost complete stop of the feeding in the acidic stream	Dangles and Guérold (2000)

Table 2 (continued)

Species	Test substance/media	Type of exposure	Exposure period	Endpoints/ biomarkers	Effect/effect concentration	Remarks	Reference
G. pulex physically pre-disrupted precopula pairs	LD high: (0.5–2.0 mg/L); low: (0.5–5.0 μg/L)	Aqueous exposure, static with conditioned leaf discs	High: 2–20 min + 24 hour post-exposure; low: 48 hour	Feeding rate, precopula re-pairing	Significantly reduced feeding activity: low: 5.0 μg/L LD; high: 1 and 2 mg/L LD	Re-pairing affected by treatments combining higher concentrations and longer exposures	Malbouisson et al. (1995)
E. toletanus	Unionized ammonia (NH$_3$-N): 0.06–0.33 mg/L NH$_3$-N	Static aqueous exposure with conditioned leaf discs	6 days	Egestion rate (mg dry weight feces/mg dry weight amphipod/day)	After 6 days at 0.3 mg/L NH$_3$-N significantly lower egestion rate. NOEC (no observable effect level): 0.18 mg/L NH$_3$-N	NOEC confirms the calculated safe concentration of 0.14 mg/L NH$_3$-N	Alonso and Camargo (2004)
G. pulex, A. aquaticus (2-week-old males)	Water from upstream and downstream of landfill leachate discharge site	Comparison of in situ and ex situ (static renewal) feeding test	6 days	Mortality, feeding rate	Mortalities and feeding rates had similar trend during in situ and ex situ exposure, but responses were amplified in situ.	In situ toxicity tests are a more precise monitoring technique	Bloor and Banks (2006)

Table 2 (continued)

Species	Test substance/media	Type of exposure	Exposure period	Endpoints/biomarkers	Effect/effect concentration	Remarks	Reference
G. pulex from two different streams, unpolluted and metal polluted	In situ: metalliferous effluents. Laboratory: Iron (Fe): 1, 2, or 3 mg/L Fe Manganese (Mn): 0.1, 0.3, 0.5 mg/L Mn	In situ exposure and lab experiment to validate field data	6 days	Feeding rate, mortality	In situ: significant higher mortality in *G. pulex* from unpolluted stream. Significantly reduced feeding rate in unpolluted *G. pulex* at metal-polluted sites. *Laboratory*: significant reductions in feeding rate at >2.0 mg/L Fe for both populations. Significant mortality increase	Animals from metal-polluted sites might be less sensitive than those from unpolluted sites	Maltby and Crane (1994)
G. pulex adult males > 5 mm	Zinc (Zn), linear alkylbenzene sulfonate (LAS), LD: pirimiphos-methyl (PM), permethrin (P)	Aqueous static renewal	Feeding inhibition: 144 hour Biomarkers: 24 and 48 hour	Feeding inhibition; biomarkers: cholinesterase (ChE), glutathione-S-transferase (GST)	ChE: significant reduction: 24 hour: 0.77 μg/L PM 48 hour:1.92 μg/L PM GST: significant increases; 24 hour: 12.3 μg/L LD 48 hour: 6.14 μg/L LD; 0.12 μg/L P	Significant reduction in feeding rate (144 hour): PM: 0.049 μg/L; P: 0.009 μg/L; LD: 3.70 μg/L; Zn: 0.04 mg/L; LAS: 0.062 mg/L	McLoughlin et al. (2000)

Table 2 (continued)

Species	Test substance/media	Type of exposure	Exposure period	Endpoints/biomarkers	Effect/effect concentration	Remarks	Reference
G. pulex	4-Nonylphenol (NP) leaf disc: 100 μg/g NP, for aqueous exposure in net	Comparison of aqueous uptake and feeding uptake	48 hour	Feeding rate, uptake depuration	Higher NP body burden from dietary exposure, but uptake from aqueous exposure unexpectedly high; predominant uptake root	No difference in depuration rates after aqueous and dietary exposure	Gross-Sorokin et al. (2003)
G. pulex	Cadmium (Cd) (2.1 and 6 μg/L)	Static aqueous exposure	24-hour exposure + 24-hour post-exposure	Post-exposure feeding depression	Significant decrease in post-exposure feeding rate at 6 μg/L Cd		Brown and Pascoe (1989)
G. pulex parasite-free and parasitized	Cadmium (Cd) (2.1 and 6 μg/L)	Static aqueous exposure	24-hour exposure + 24-hour post-exposure	Post-exposure feeding depression Mortality	Parasitized G. pulex significant higher mortality to 2.1 μg/L Cd than uninfected	Parasitized consumed only 17–21% of the food eaten by uninfected G. pulex when held in dilution water	Brown and Pascoe (1989)

Table 2 (continued)

Species	Test substance/media	Type of exposure	Exposure period	Endpoints/biomarkers	Effect/effect concentration	Remarks	Reference
G. pulex, A. aquaticus	LD, DCA	Species interactions in toxicant systems	96, 240 hour	Feeding activity (FT_{50}) on A. salina eggs	LD: significantly reduced FT_{50} of G. pulex coexposed with A. aquaticus; 96 hour: >3.8 µg/L 240 hour: > 6.5 µg/L DCA 90 µg/L: 100 and 60% survival for A. aquaticus and G. pulex, respectively. G. pulex no longer dominant species	Exposure to low LD concentrations (0.1 and 0.9 µg/L): significant increase in gammarid feeding activity	Blockwell et al. (1998)
G. pulex males 7–9 mm	Cadmium chloride (CdCl) (7.5 and 15 µg/L)	Physiological and behavioral responses, static renewal	168 hour	Feeding rate, hemolymph Ca^{2+} conc., osmolality Na^+/Cl^- conc., Cd accumulation Na^+/K^+-ATPase, locomotion, Ventilation	Significant decrease in osmolality, hemolymph Ca^{2+} conc., mortality, feeding rate, locomotor and ventilatory activities but not hemolymph Na^+/Cl^-	Significant increase in Na^+/K^+-ATPase activity	Felten et al. (2008)

to beakers containing 18 mL of the relevant toxicant or control solution, together with 10 shell-less eggs of *A. salina*. The number of eggs eaten in each beaker is recorded frequently, which allows median feeding times (FT_{50}, the time at which 50% of the eggs have been consumed) to be determined. This nondestructive method provides a rapid indication of the status of groups of individuals (Taylor et al. 1993). Blockwell et al. (1998) used this feeding assay to investigate effects on juvenile *G. pulex* exposed to the freshwater pollutants copper, lindane, and 3,4-dichloroaniline (3,4-DCA). Gammarid feeding was reduced following 96-hour exposure at 12.1 μg/L copper or 8.4 μg/L lindane, and following 240-hour exposure at 918 μg/L of 3,4-DCA. A sustained reduction in feeding rates may cause growth inhibition and impaired reproduction. This has been previously identified as sublethal responses of other freshwater organisms exposed to comparable concentrations of lindane, 3,4-DCA, or copper (Taylor et al. 1991). Interestingly, the feeding rate was higher for *G. pulex* specimens that had been exposed for 240 hour to 0.09 μg/L lindane, when compared to controls; possibly this was caused by stimulatory effects associated with lindane at low exposure concentrations (Blockwell et al. 1998).

4.2 Food Choice Experiments

Feeding experiments may be enhanced by including food choice as an endpoint. Bundschuh et al. (2009) used such an experimental setup to assess whether an antibiotic mixture, consisting of sulfamethoxazole, trimethoprim, erythromycin-H_2O, roxithromycin, and clarithromycin, had an effect on aquatic communities comprising bacterial and fungal decomposers and invertebrate detritivores. Two types of leaf discs were offered to an amphipod shredder, *G. fossarum*, after those discs had been conditioned with and without antibiotics (total concentration of 2 or 200 μg/L) for approximately 20 days. Interestingly, *G. fossarum* preferred the antibiotic-conditioned leaf discs at 200 μg/L over the control discs (pair-wise *t* test; $p = 0.006$). A similar but not significant tendency was observed for leaves conditioned with antibiotics at 2 μg/L. The number of bacteria associated with leaves did not differ between treatments at either antibiotic concentration (*t* test; $p = 0.57$), but the fungal biomass (measured as ergosterol) was significantly higher in the 200 μg/L treatment (*t* test; $p = 0.038$), suggesting that the preference of *Gammarus* may be related to a shift in fungal communities. This may indicate that mixtures of antibiotics may disrupt important ecosystem processes such as organic matter flow in stream ecosystems (Bundschuh et al. 2009).

The importance of fungi in the diet of *G. pulex* and *A. aquaticus* was also stressed by another study (Graça et al. 1993). Herein, in feeding trials with fungal mycelia, either fungally colonized or fungally uncolonized leaves were used to assess feeding preferences. The authors found that *A. aquaticus* scrapes at the leaf surface and selectively consumes fungal mycelia, whereas *G. pulex* nibbles leaf material, apparently only for the quality of the leaf tissue, not for any microorganisms present. This result was reinforced experimentally: *G. pulex* ignored pure fungal mycelia and preferentially consumed conditioned and, to a lesser extent, unconditioned leaf discs. Food choice of *A. aquaticus* was positively correlated with fungal biomass, whereas for *G. pulex,* fungi appear to be more important as modifiers of leaf material (Graça et al. 1993). Notwithstanding, both gammarids exhibited strong preferences for *Anguillospora longissima* and *Heliscus lugdunensis* (Graça et al. 1994). The relative abundance of the fungi was not important. However, the authors observed intra-population variability in food preferences, both between individuals and for the same individual through time. When looking at different stream categories, microbial breakdown rates of oak and alder leaves did not differ, although aquatic hyphomycete-species richness on leaf litter positively correlated with riparian plant-species richness (Lecerf et al. 2005). Fungal-species richness may enhance the breakdown rate of leaf litter through positive effects on resource quality for shredders; a feeding experiment confirmed this positive relationship between fungal-species richness per se and leaf litter consumption rate by the amphipod shredder *G. fossarum.* Hence, plant-species richness may indirectly govern ecosystem functioning through complex trophic interactions (Lecerf et al. 2005). Integrating microbial diversity and trophic dynamics in in situ exposures of gammarids may improve the understanding and assessment of aquatic pollution.

4.3 Leaf-Mass Feeding Assays Linked to Food Consumption

Dangles and Guérold (2000) conducted a *G. fossarum* feeding assay with emphasis on consumption of leaf mass, a parameter that could be used to assess potential changes in ecosystem processes. This approach allows one to measure if, for example, acidification or another external impact on a stream leads to alteration in leaf breakdown, which may affect accumulation of leaf litter. Such an effect can be assessed by comparing leaf dry weight and area between undisturbed/unexposed and disturbed/exposed sampling areas. Dangles and Guérold (2000) used beech leaves and conditioned them for 136 days, either in a highly acidic stream (mean pH 4.68; mean alkalinity −19 µEq/L; mean total Al 801 µg/L) or in a reference stream (mean pH 7.36; mean alkalinity 539 µEq/L; mean total Al 36 µg/L). The two diets were tested in situ with adult males in feeding trial units that were then placed in tanks anchored to stream bottoms. After 6 days, the organisms were removed, gammarids and leaf disks were dried, weighed, and the consumed leaf mass determined (Dangles and Guérold 2000). No mortalities were observed, but *G. fossarum* almost completely ceased feeding activity in the acidic stream, and after 6 days, feeding activities on both exposed and control leaves were 113 and 46 times lower than in the reference stream. This experimental setup nicely shows that the water quality of the stream has an impact on the feeding activity of gammarids. Under acidic-stress conditions, gammarids are apparently unable to increase or even maintain their energy uptake from available food resources. In addition, the quality of the leaves had an impact on the feeding activity of gammarids (Dangles and Guérold 2000).

4.3.1 Feeding Activity and Survival Related to Toxicity or Abiotic Parameters (Ex Situ)

The effects of toxicants on gammarid feeding activity and survival were examined in several studies (Alonso and Camargo 2004; Bloor and Banks 2006; Maltby and Crane 1994), and in one study McLoughlin et al. (2000) investigated biochemical biomarkers in addition to feeding activity and survival.

Alonso and Camargo (2004) investigated the toxicity of unionized ammonia (NH_3) to determine if the presence of this more toxic form of ammonia has a negative effect on the feeding activity of amphipods. Instead of leaf-disk-measurements, the authors measured egestion rate (milligram dry weight of feces/milligram dry weight of amphipod/day) as an endpoint for feeding activity. Feeding activity was investigated during a 6-day exposure of the amphipod *Eulimnogammarus toletanus* to 0.06–0.33 mg/L NH_3-N, by measuring the egestion rate on days 2 and 6 of exposure. After 2 days of exposure, the egestion rate was slightly higher in exposed compared to control animals, but the difference was not significant. After 6 days of exposure, however, the egestion rate of *E. toletanus* exposed to 0.3 mg/L NH_3-N was significantly lower than for the control. The NOEC (no observable effect concentration), based on the second assay, was 0.18 mg/L NH_3-N, confirming the calculated safe concentration of 0.14 mg/L NH_3-N (Alonso and Camargo 2004).

Bloor and Banks (2006) recently evaluated mixed species feeding assays with the pollution-sensitive *G. pulex,* and the pollution-tolerant *A. aquaticus*, to determine if test animals' response was comparable during in situ vs. ex situ toxicity tests. Seven sampling points were chosen along a stream, which received leachate discharge from an unlined, unused UK landfill site. Points A and B were upstream of the contamination, C was adjacent to the influx, and D–G were downstream of the leachate discharge. For the in situ and ex situ tests, 2-week-old male laboratory-bred *A. aquaticus* and *G. pulex* were used as test animals and were either (1) transplanted to the seven sampling points for the duration of the in situ tests or (2) exposed in the laboratory (in situ) to water samples taken from each sampling site. The authors found that mortalities and feeding rates followed similar trends during the in situ and ex situ tests, but that the response of test animals was amplified during in situ testing. The authors also observed that the effects were greater in April than in August, possibly because of a higher rainfall frequency during the spring, which may have resulted in greater portions of flushed contaminant load from the landfill site and thus higher levels of pollution leaching into the stream. The conclusion, therefore, was that in situ toxicity tests are a more precise monitoring technique, when compared to ex situ assays (Bloor and Banks 2006).

A similar study was performed a few years earlier to evaluate whether an in situ bioassay was useful for evaluating the impact of complex effluents on freshwaters and to identify toxic components (Maltby and Crane 1994). In this study, *G. pulex* was exposed to metalliferous effluents that reduced the feeding rate, which was a sensitive indicator. Chemical analysis showed that the effluents contained a mixture of five potentially toxic metals. Based on feeding rate and bioaccumulation data, two metals (iron and manganese) were identified as the probable toxic agents.

Laboratory experiments were performed to validate the results of the field conclusions. Results confirmed that iron was a major toxicant. During both the field and laboratory studies, the sensitivities of *G. pulex* from a metal-contaminated and a clean site were compared. Interpopulation response differences to the toxicants for *G. pulex* were found in the field study, but not in the laboratory experiments; this indicates that animals from metal-polluted sites may be less sensitive than those from unpolluted sites.

4.3.2 Feeding Rate, Uptake, and Depuration

In general, the major exposure and uptake route for soluble toxins by aquatic organisms is regarded to be through the water column. For hydrophobic chemicals, however, exposure and uptake through the diet is often of greater importance, because the chemicals adsorb onto organic sediments and food. To date, the relative importance of water and food in the uptake and accumulation of a toxin by a benthic detritivore has not been assessed. Gross-Sorokin et al. (2003) investigated the importance of feeding rate on uptake and depuration of the hydrophobic endocrine disruptor 4-nonylphenol (NP) by *G. pulex*. These authors compared this result with NP uptake from leaves into the water. To this end, dried horse chestnut leaf disks were soaked in stream water and then dosed with NP for 24 hour to achieve a nominal, environmentally relevant leaf concentration of 100 μg/g NP. NP concentrations were measured in the gammarids, water, and leaf disks at 8, 12, 24, and 48 hour, respectively. In addition, the amount of leaf material ingested relative to the total weight of the gammarids was measured. After 48 hour of exposure, gammarids were transferred to clean water to determine the degree of depuration. By using a bootstrap nonlinear regression technique, the authors showed that a higher body burden of NP resulted from dietary exposure, partly because of the higher source concentration in leaf disks. If leaf-disk concentration is taken into account, uptake from water is unexpectedly high and confirms the assumption that it is the predominant uptake route and derives from the large volumes transferred across gills. Depuration rates following aqueous and dietary exposures were rapid and did not differ significantly among different treatments (Gross-Sorokin et al. 2003).

4.4 Modeling of Feeding Activity and Rate

4.4.1 Field Estimates of Feeding Rates – Modeling, Ingestion, and Egestion Rates

Attempts were made to understand and estimate field-feeding rates by using exponential regression models (Marchant and Hynes 1981). The feeding rate of *G. pseudolimnaeus* was measured monthly for 7 months in the field by monitoring the weight decline of gut contents, when the amphipod was starved. This weight decline was then modeled by an exponential regression of weight on time. With the assumption that amphipods are continuous feeders, feeding rate was calculated by multiplying the dry weight of a full gut by the specific rate of gut emptying, i.e., the slope of the exponential regression. The authors found that the specific rate of emptying was independent of animal size but increased with temperature, resulting in longer digestion times at lower temperatures. This may lead to an increase in assimilation efficiency at lower temperatures. In the laboratory, however, the assimilation efficiency of amphipods that fed on decaying maple leaves was only 10% and did not vary with temperature. The turnover time of the contents of a full gut, however, was often similar to the turnover time measured in the field, i.e., the reciprocal of the specific rate of emptying, thus confirming the use of an exponential regression model (Marchant and Hynes 1981).

4.4.2 Carnivorous Feeding Activity

Werner et al. (2002) used a similar approach to estimate the potential predation impact of the under-ice arctic amphipod *G. wilkitzkii* by combining information on ingestion rates with population densities. The carnivorous feeding activity and the energy budget were studied by estimating a maximum potential ingestion rate (I_{max}). This I_{max} value was 2.1±0.4% of body carbon/day and was calculated from an allometric literature function and from body mass. In addition, respiration measurements were integrated to assess the lower specific ingestion rates (1.4±0.4%) of body carbon/day, required to meet metabolic demands. Feeding experiments, with co-occurring pelagic calanoid (*Calanus* spp.) or sympagic harpacticoid (*Halectinosoma* sp.) copepods as prey, were conducted. From these prey species, actual ingestion rates of 8.0±5.6% of body carbon/day and 0.1±0.1% of body carbon/day, respectively, were determined. The values indicate that predatory feeding on pelagic copepods may constitute an important food source for *G. wilkitzkii*. Estimates on the potential predation impact of *G. wilkitzkii* were made by combining information on ingestion rates with population densities. Predation impact was very high on *Calanus* spp. in the under-ice water layer (61.5% of under-ice standing stock/day), but comparatively low on *Halectinosoma* sp. in the bottom of the ice (3.8% of standing stock/day) (Werner et al. 2002). The observation that *G. wilkitzkii* prey on pelagic copepods is obviously significant and potential predation impact should be taken into account when studying freshwater amphipods.

4.5 Post-exposure Feeding Depression Assay

Recently, a bioassay that uses post-exposure feeding depression (PFD) in *D. magna* as an endpoint has been developed to assess toxic effluent in rivers (McWilliam and Baird 2002b). Previous studies have suggested that *D. magna*, exposed to toxic substances, may exhibit delayed feeding behavior recovery, suggesting that PFD is a sensitive and robust endpoint (McWilliam and Baird 2002b) that could be included in toxicity assessments with *G. pulex*.

McWilliam and Braid (2002a) investigated whether this endpoint could be used in the field as an in situ bioassay with *D. magna;* in this study, *D. magna* were exposed in test chambers to four known or suspected contaminated and reference sites. The authors found that the bioassay was reliable for use in the field, because more than 90% post-exposure survival of test organisms was observed and allowed post-exposure feeding rates to be measured. At each contaminated site, significant depression in post-exposure feeding rates was recorded (McWilliam and Baird 2002a). Feeding depression was also observed in a similar study conducted with *G. pulex*, wherein 6 µg/L of cadmium produced a decreased feeding rate after a 24-hour exposure (Brown and Pascoe 1989). Although depression in post-exposure feeding rates was apparent at all but one contaminated site in the McWilliam and Braid study, no impact was detected on the benthic macroinvertebrate community when using the Biological Monitoring Working Party scoring system (McWilliam and Baird 2002b). This indicates that post-exposure feeding depression was a more reliable and sensitive endpoint for detecting toxicity than were changes in community structure and therefore offers a sensitive, robust, and ecologically relevant diagnostic endpoint for use in water quality assessment schemes.

4.6 Effects of Parasites on Gammarid Feeding Ecology

The amphipod *G. pulex* is an intermediate host to the acanthocephalan fish parasite *Echinorhynchus truttae* (Fielding et al. 2003). It is not clear if feeding activity (herbivore, detritivore, or predatory) of *G. pulex* is affected when it is parasitized by *E. truttae*. Fielding et al. (2003) investigated the effects of *E. truttae* parasitism on the following components of the *G. pulex* diet: stream-conditioned leaves, dead chironomids, and the live juvenile isopod *A. aquaticus*. After 21 days of exposure, parasitism had no effect on daily feeding rates or wet weights of *G. pulex* that fed on leaves or chironomids. However, parasitism significantly affected the number of *A. aquaticus* killed by *G. pulex*, with parasitized individuals killing significantly fewer and smaller *A. aquaticus* than did their unparasitized counterparts.

The above findings on *G. pulex* raise the question if parasite-related behavioral changes also negatively affect the response of *G. pulex* toward toxicant exposure (Brown and Pascoe 1989). To test this question a total of 600 *G. pulex* were collected from the River Teme and 52% of the collected population were infected with *Pomphorhynchus laevis*; 20% contained larvae, and 80% contained cystacanths, the stage that is infective to fish, the definitive host. After 80 days of exposure to dilution water, 14% of uninfected and 94% of infected *G. pulex* were dead. When infected *G. pulex* were exposed to a nominal cadmium concentration of 2.1 µg/L, they were significantly more sensitive (LT_{50} = 17.5 days) to the toxicant than were uninfected conspecifics (LT_{50} = 42 days), while at 6 µg/L, differences in mortality rates between infected and uninfected individuals were insignificant (Brown and Pascoe 1989). In general, the authors found that infected *G. pulex* consumed only 17–21% of the food eaten by uninfected *G. pulex*, when held in dilution water. A further disadvantage for *G. pulex* was that *P. laevis* was insensitive to cadmium (Brown and Pascoe 1989).

5 Behavior

Behavioral assays are used to detect several individual and inter-individual behaviors, i.e., avoidance or preference reactions of macroinvertebrates to toxic pollutants. Table 3 gives an overview of different behavioral tests conducted with gammarids. These tests, or their endpoints, originate from antipredator behaviors such as drift, reduction of activity, habitat choice, and diel pattern of movement.

5.1 Antipredator Behavior

5.1.1 Drift Behavior Linked to Predators

It is known that drift of *Gammarus* individuals was reduced by chemical clues remaining from the presence of predatory fish (Andersson et al. 1986; Friberg et al. 1994; McIntosh et al. 1999). Predation on amphipods in leaf litter was found to be significantly lower than in other microhabitats, showing the importance of habitat choice as an antipredator behavior (Holomuzki and Hoyle 1990). To respond to the fluctuating risk of the presence of predator, prey need information on predation risk, and the quality of this information is reflected in prey behavioral time lags. For example, prey benefit from remaining in a refuge long after the predator has left. Experimental investigation showed that prey behavior time lags were longer when predator density was higher and prey were less hungry, and had lower escape ability (Sih 1992).

Wisenden et al. (1999) examined whether *G. minus* exhibit antipredator behavior in response to injury-released chemicals from conspecifics or heterospecifics (Crustacea and Isopoda). Relative to the control group, *G. minus* exposed to conspecific cues showed decreased activity and moved to the substratum after detecting the cue. Alternatively, *G. minus* moved up into the water column and increased activity after being exposed to the heterospecific cue. In addition, the time to first attack increased when *G. minus* was exposed to a predator (green sunfish) if *G. minus* was also exposed to conspecific cues; in contrast, the time to first attack decreased when *G. minus* were exposed to heterospecific cues. Thus, *G. minus* seems to benefit from its antipredator response to conspecific cues (Wisenden et al. 1999).

5.1.2 Drift Behavior Resulting from Pollution

Antipredator behavior may be induced not only by a predator but also by chemical threads. Previous studies on pulsed exposures to pyrethroid insecticides showed that many stream macroinvertebrates respond to such exposures by demonstrating catastrophic drift (Breneman and Pontasch 1994; Sibley et al. 1991). Lauridsen and Friberg (2005) investigated the behavioral changes of stream macroinvertebrates exposed to a pulse of the pyrethroid insecticide lambda-cyhalothrin in outdoor experimental channels. The number of macroinvertebrates drifting, as well as the

Table 3 Test methods with *gammarids* to assess toxicant effects on behavior

Species	Test substance/ media	Exposure type	Exposure period	Endpoints/ biomarkers	Effect/effect concentration	Remarks	Reference
G. pulex, B. rhodani, L. fusca	Pyrethroid insecticide lambda-cyhalothrin (LC): 0.001, 0.01, 0.1, 1.0 μg/L	Pulsed exposure in outdoor experimental channels	2 hour + 60 min pulse + 24 hour	Drift response	*G. pulex*: > 0.001 μg/L LC, significant catastrophic drift. *B. rhodani, L. fusca*: > 0.01 μg/L LC, significant catastrophic drift	Drift behavior of insect species (*B. rhodani, L. fusca*) is less sensitive than that of *G. pulex*	Lauridsen and Friberg (2005)
G. pulex	Unpolluted stream, copper (Cu)-polluted stream (peak concentrations: 69, 58, 30 μg/L Cu-tot)	In situ: one reference + one local population. One simulated pulsed Cu exposure: 70 μg/L Cu^{2+}	5–6 days: 24-second recordings every 30 min	Locomotion and ventilation activity with the MFB®	Reference population was significantly less active than local population. Cu-pollution pulses increased number of active organisms and time spent in locomotion	No effects on natural drift behavior	Gerhardt et al. (1998)

Table 3 (continued)

Species	Test substance/ media	Exposure type	Exposure period	Endpoints/ biomarkers	Effect/effect concentration	Remarks	Reference
G. pulex (8–10 mm)	Pharmaceuticals: fluoxetine: 0.1–100,000 ng/L; ibuprofen: 1–100,000 ng/L; carbamazepine: 1–100,000 ng/L. Surfactant: cetyltrimethylammonium bromide (CTAB): 0.0001–100 mg/L	Short-term exposure with static renewal	1.5 hour: 240-second recordings every 10 min	Activity pattern: locomotion and resting activity with the MFB®	10–100 ng/L fluoxetine and ibuprofen: significant activity reduction. 10–100 ng/L CTAB: acute immobilization	Observed behavioral effect concentrations for all three chemicals were 104–107 times lower than previously reported LOECs (lowest observable effect concentrations), in the range of environmentally concentrations	De Lange et al. (2006a)
G. pulex (5–8 mm)	Rivers: Aller (clean water) Meuse (simulated pulses of metals and organic xenobiotics) Rhine (frequent natural pulsed pollutions)	Aller: in situ biomonitoring; Meuse: water flow-through, ex situ, +/− simulated pulsed exposures; Rhine: in situ	4-min recordings every 10 min Aller: 14 days; Meuse: 5 days; Rhine: 1.5 months	Locomotory activity with the MFB®	Aller: no effect; Meuse: 20% decrease in activity (1st pulse) and increased mortality (2nd, 3rd pulse); Rhine: 20% activity decrease during biomonitoring	G. pulex in the MFB is a suitable alert system for water quality monitoring at sensitive sites and sites with accidental pollution	Gerhardt et al. (2007)

Table 3 (continued)

Species	Test substance/ media	Exposure type	Exposure period	Endpoints/ biomarkers	Effect/effect concentration	Remarks	Reference
E. meridionalis; H. pellucidula; C. picteti	Acid mine drainage (AMD)	Simulated short-term acid pulses	40 hour, with a 1-hour pulse of AMD after approx. 12 hour	Locomotory and ventilatory activity with the MFB ®	E. meridionalis: increased locomotion, then increased ventilation. C. picteti: increased locomotion H. pellucidula: no effect	Sensitivity: E. meridionalis > C. picteti > H. pellucidula	Macedo-Sousa et al. (2008)
G. duebeni (15–21 mm)	Cu Pentachlorophenol (PCP), benzo[a]pyrene (B[a]P)	48 hour static renewal	7 days	Pleopod beat frequency, swimming endurance	Significantly impaired swimming endurance at 45 μg/L Cu, 20 μg/L PCP, 8 μg/L B[a]P. Pleopod beat frequency: 50 μg/L Cu	Swimming endurance is the more sensitive and consistent endpoint	Lawrence and Poulter (1998)
C. marinus (15–21 mm)	Cu PCP B[a]P	Flow-through	180 hour	Swimming endurance	Significantly impaired swimming endurance at 15 μg/L Cu, 40 μg/L PCP, 20 μg/L B[a]P		Lawrence and Poulter (2001)

mobility of macroinvertebrates caught in the drift, was assessed. In small replicated subsections of the experimental channels, two insect species (*Baetis rhodani, Leuctra fusca/L. digitata*) and an amphipod (*G. pulex*) were allowed to acclimatize for 26 hour before 60-min pulsed exposures to lambda-cyhalothrin (0.001, 0.01, 0.1, and 1.0 µg/L). Measurement started 2 hour before the insecticide was applied and continued for the following 24 hour. All three species responded to the insecticide pulse by demonstrating catastrophic drift, starting at 0.001 µg/L for *Gammarus*, whereas the drift response threshold was 0.01 µg/L for the two insect species (Lauridsen and Friberg 2005). The higher the pulse concentration, the more the individuals of the three species exhibited drifting behavior. Drift response onset was directly linked to the applied pulse concentration, with the highest concentrations resulting in more individuals of all species entering drift at an early stage. In addition, both insect species were in the process of being immobilized at the two highest exposure concentration; in *Gammarus*, this was only the case for a few individuals. The clear species-specific responses indicate that sublethal doses have the potential to change macroinvertebrate community structure.

5.1.3 The Effect of Parasitism on Antipredator Behavior

Baldauf et al. (2007) examined whether a parasite infection (*P. laevis*) influenced an amphipod's reaction to a fish predator's odor. A series of choice experiments, with infected and uninfected *G. pulex*, were performed to distinguish between the effects of visual and olfactory predator cues on parasite-induced changes in host behavior. The authors reported that uninfected individuals significantly avoided predator odors, while infected individuals significantly preferred the side with predator odors (Baldauf et al. 2007). However, when only visual contact with a predator was allowed, infected and uninfected gammarids behaved similarly and had no significant preference, indicating that the parasite *P. laevis* increases its chance of reaching a final host by olfactory-triggered manipulation of the antipredator behavior of its intermediate host.

A similar study was conducted on parasite-induced behavior and color changes in *G. pulex* and its relevance to the risk of predation by fish (Bakker et al. 1997). In infected *G. pulex*, the conspicuous orange yellow parasite *P. laevis* is visible through the transparent cuticle of *G. pulex*. Because it was previously reported that infected gammarids are significantly less photophobic than uninfected ones (Cezilly et al. 2000), Bakker et al. (1997) tested whether parasite color and parasite-induced changes in host behavior affected the predation rate of *G. pulex*. Therefore, hungry three-spined sticklebacks (*Gasterosteus aculeatus*) were offered uninfected *G. pulex* that had a painted orange spot on their cuticle (to simulate infection) and infected *G. pulex* with brown paint on the parasite infection spot. Sticklebacks consumed significantly more infected *G. pulex* than uninfected ones. Experimental exclusion of behavior or color effects of the parasite on its intermediate host showed that both parasite color and parasite-induced changes in *G. pulex* behavior significantly increased their vulnerability to predation by sticklebacks.

The joint influence of two acanthocephalan parasites on the behavior of *Gammarus* has also been investigated (Cezilly et al. 2000). These authors studied the effects of simultaneous infection by a fish acanthocephalan parasite (*P. laevis*) and a bird acanthocephalan parasite (*Polymorphus minutus*) on the behavior of their common intermediate host, *G. pulex*. In this study, the reaction to light and vertical distribution of infected (by one or both parasites) and uninfected individuals was investigated. Generally, uninfected gammarids tended to be at the bottom of the water column and were photophobic, but *P. laevis*-infected gammarids were attracted to light; *P. minutus*-infected individuals showed a modified vertical distribution and swam closer to the water surface (Cezilly et al. 2000). The observed behavioral changes of the organisms infected with both parasites seemed to be dependent on only the presence but not the intensity of the parasite infection. Gammarids harboring both parasites were vertically half-way between those of *P. laevis*- and *P. minutus*-infected individuals, whereas *P. laevis* was able to induce altered reaction to light even in the presence of *P. minutus*.

5.2 Multispecies Freshwater Biomonitor® (MFB)

5.2.1 Method Description

Gerhardt et al. (1994) developed an online biomonitoring system capable of quantitatively assessing several aspects of behavior in situ and ex situ, and doing so in real time. The system is based on a quadrupole impedance conversion technique that simultaneously records several behavioral parameters of a wide range of aquatic organisms, such as *D. magna, G. pulex, Sialis lutaria, Leptophlebia vespertina, B. niger,* Simuliidae, *Dinocras cephalotes, Hydropsyche siltalai,* and tadpoles of *Rana temporaria*. During exposure, the organisms move freely between two pairs of electrodes on each sidewall of a test chamber, which receives unfiltered stream water or exposure water (Gerhardt et al. 1994). The organism's behavior is expressed as movements that lead to changes in an electrical field and these are measured as changes in the impedance of the system. For example: (1) locomotion: swimming and crawling result in irregular amplitudes and frequencies, (2) resting: small signals that cannot be separated from background noise, (3) ventilation: regular, high-frequency movements with, e.g., pleopods to establish a constant water flow across the gills, (4) feeding: species-specific patterns for grazing, filtering, and hunting. The impedance converter proved to be a sensitive and quantitative tool for use in behavioral, ecological, and ecotoxicological studies, which makes it a promising tool for continuous biomonitoring purposes (Gerhardt 1999).

5.2.2 Behavioral Changes in the MFB Related to Toxic Effects

The biomonitoring system was used to compare the behavior of two different *G. pulex* populations, one originated from an anthropogenically unpolluted stream and the other from a copper-polluted study site, where the biomonitor was placed (Gerhardt et al. 1998). *G. pulex* were exposed to simulated copper pollution peaks of 70 µg Cu^{2+}/L and their drift behavior was compared to natural drift. A nocturnal drift peak was observed for both *G. pulex* populations. The exposure resulted in significantly less activity (number of active organisms per day and time spent in locomotion and ventilation) among members of the reference population compared to the local population. Copper pollution pulses provoked increased activity in a number of organisms that were active in both populations and the time these organisms spent on locomotion. However, no significant changes in the natural drift were registered, probably from dilution downstream of the pulse (Gerhardt et al. 1998).

DeLange et al. (2006a) also used the MFB® to investigate whether a prolonged exposure to low concentrations of anthropogenic chemicals may lead to sublethal effects, including changes in behavior. In their study, *G. pulex* were exposed to three pharmaceuticals, the antidepressant fluoxetine, the analgesic ibuprofen, and the anti-epileptic carbamazepine, and one cationic surfactant, cetyltrimethylammonium bromide (CTAB). Low concentrations (10–100 ng/L) of fluoxetine and ibuprofen resulted in a significant decrease in activity, and the response to carbamazepine showed a similar pattern; however, differences were not significant. The surfactant CTAB led to a dose-dependent decrease in activity with increasing concentrations. Surprisingly, the observed behavioral effect concentrations for all three chemicals were 104–107 times lower than previously reported LOECs (lowest observable effect concentrations) and were in the range of actual environmental concentrations (De Lange et al. 2006a).

Another in situ biomonitoring study was conducted along the rivers Meuse (NL), Aller (GER), and Rhine (F) for the purpose of validating the MFB® for an in situ application approach with *G. pulex* as the new indicator species (Gerhardt et al. 2007). Three field sites were selected for characteristics adequate to answer the following research questions: (1) Is *G. pulex* able to survive in clean unfiltered surface water (Aller River), with detritus as the food source in the MFB? (2) Is *G. pulex* able to react to a cocktail of metals or organic xenobiotics that are applied as pulse pollution in concentrations relevant to those occurring under accidental circumstances (Meuse River)? (3) Is the MFB with *G. pulex* ready to be used for a long-term evaluation at a location (Rhine River), with frequent pulse pollution and changes in water quality? *G. pulex* used in the reference stream Aller did not show any negative effects and had a 100% survival. Alternatively, *G. pulex* responded to a pulsed exposure of a mixture either of trace metals or of several organic xenobiotics, with up to a 20% decrease in locomotory activity (at the 1st pulse) and increased mortality (at 2nd or 3rd pulse only). Deployment at the monitoring station on the Rhine River demonstrated that *G. pulex* were able to detect chemical irregularities by displaying up to a 20% decrease in locomotory activity, confirming the suitability of *G. pulex* in the MFB as an alert system for water quality monitoring at sensitive or accidentally polluted sites (Gerhardt et al. 2007).

5.2.3 Behavioral Early Warning Responses After Pulsed Exposures

The effects of draining abandoned mines were investigated by exposing *E. meridionalis, H. pellucidula,* and *Choroterpes picteti* to short-term pulses of drained acid mine waste (Macedo-Sousa et al. 2008). Possible negative effects of such pulses, resulting from acidity and heavy metal contamination, were assessed by using the MFB® behavioral early warning responses (locomotion and ventilation). *E. meridionalis* was the most sensitive species in terms of mortality and behavioral endpoints, followed by *C. picteti* and *H. pellucidula*; this demonstrates the suitability of using benthic invertebrates' behavioral early warning responses for detecting spikes of pollutants. Exposed *E. meridionalis* showed increased locomotion, with a subsequent increase in ventilation, whereas *C. picteti* reacted with increased locomotion; *H. pellucidula* was unaffected.

5.3 A Sublethal Pollution Bioassay with Pleopod Beat Frequency and Swimming Endurance

Lawrence and Poulter (1998) developed a bioassay with *G. duebeni* for sublethal ecotoxicological studies, based on the monitoring of pleopod beat frequency (ventilation) and swimming endurance against a head flow of water. Pleopod beat frequency showed a complex dose- and time-dependent response to copper, whereas swimming endurance showed a clear dose–response relationship and was identified as the more reproducible technique (Lawrence and Poulter 1998).

By using swimming efficiency again, Lawrence and Poulter (2001) investigated the effect of copper, pentachlorophenol (PCP), and benzo[a]pyrene (B[a]P) on the amphipod *Chaetogammarus marinus*. Moreover, swimming endurance was determined by following the protocol developed by Lawrence and Poulter (1998). Swimming endurance was significantly impaired at concentrations of 15 µg/L copper, 40 µg/L PCP, and 20 µg/L B[a]P.

5.4 Behavior in Combination with Other Endpoints

5.4.1 Drift and Foraging Activity

Behavioral aspects, combined with feeding activity, were examined by Allan and Malmqvist (1989). These authors studied the relationship between drift and foraging activity in *G. pulex* by comparing catches of *G. pulex* from the benthos, drift, and small traps baited with cheese. They investigated two field sites, one with both sculpins and trout, and one without the fish. Within 15 min, many *G. pulex* were captured in the baited traps, and a sculpin, caught in an adjacent cage, had no counteracting influence, which demonstrated the effectiveness of chemical attractants.

These authors reported that trap collections appeared to be useful for detecting small-scale spatial patterns and was an indication of a highly aggregated distribution. Traps exclusively captured organisms from the two largest size classes of *G. pulex*. In contrast, drift collections consisted almost exclusively of individuals <4 mm during the day and the larger *G. pulex* in the night drift. When analyzing stomach contents of trout and sculpins, the authors found that they selectively captured larger prey that were proportional to their size. Allan and Malmqvist (1989) attributed the rarity of larger *G. pulex* in the daytime drift to a greater daytime risk of predation, but not to the absence of foraging activity in the amphipod, because baited traps and direct observation indicated that *G. pulex* is continuously active.

5.4.2 Species Interaction and Feeding Activity in a Toxic System

Feeding bioassays may also be used to investigate species interaction in toxicant systems; in such systems, the stresses of toxicant and competition are integrated. In one such system, *G. pulex* was coexposed with *A. aquaticus* to different concentrations of lindane or 3,4-DCA, and the feeding response of *G. pulex* was recorded (Blockwell et al. 1998). A 96-hour exposure to 3.8 and 6.0 μg/L lindane led to a reduced *G. pulex* feeding rate; coexposure of *G. pulex* with *A. aquaticus* produced the same result, but at a higher rate. After 240 hour of exposure, only gammarids exposed to 6.5 μg/L lindane showed a reduced feeding rate, but exposure to very low concentrations of lindane (0.1 and 0.9 μg/L) resulted in a significant increase in gammarid feeding activity. In the 3,4-DCA coexposure of gammarids with *A. aquaticus* (96 and 240 hour), the calculation of the gammarid median feeding times (FT_{50}) could not be performed, because, in most groups, less than 50% of the *A. salina* eggs were eaten. However, a comparison to controls showed that a substantial reduction in gammarid feeding activity had occurred in the majority of the 3,4-DCA treatment groups. Interestingly, exposure to 3,4-DCA at 90 μg/L apparently reversed the direction of the species interaction, with 100 and 60% survivorship recorded for *A. aquaticus* and *G. pulex*, respectively (i.e., *G. pulex* was no longer the dominant species). The different modes of action of 3,4-DCA and lindane may be responsible for the recorded results (Blockwell et al. 1998).

5.4.3 Combined Assessment of Locomotory, Ventilatory, and Feeding Activity

In a very recent study (Felten et al. 2008), the combination of behavioral endpoints and feeding activity was used to investigate the effects of cadmium (7.5 and 15 μg/L) on physiological and behavioral responses of *G. pulex*. Mortality and whole-body cadmium concentration of exposed gammarids were found to be significantly higher than were in controls. Cadmium exposure exerted a significant decrease in osmolality and hemolymph Ca^{2+} concentration, but not in hemolymph Na^+ and Cl^- concentrations, whereas the Na^+/K^+-ATPase (adenosine triphosphatase) activity was significantly increased to maintain homeostasis. Cadmium exposure resulted in a significant reduction of behavioral responses, such as feeding rate, locomotor (number of moving animals), and ventilatory activities (pleopod beating frequency), possibly to limit energy loss and redirect it to osmoregulation and detoxification (Felten et al. 2008). The results of this study indicate that osmolality and locomotor activity in *G. pulex* could be effective ecophysiological/behavioral markers to monitor freshwater ecosystems and to assess the health of organisms and associated implications on population levels.

5.4.4 Combined Assessment of Re-pairing of Precopula Pairs and Feeding Rate

Malbouisson et al. (1995) used feeding rate in combination with re-pairing of precopulatory *G. pulex* to assess the toxicity of lindane. In this study, precopulatory pairs were physically pre-disrupted, fed with alder leaves (previously conditioned by *Cladosporium* sp.), and exposed to either high concentrations (0.5–2.0 mg/L) for 2–20 min or low concentrations (0.5–5.0 μg/L) for 48 hour at 15°C. The 48-hour exposure to 5.0 μg/L lindane led to significantly reduced feeding activity but did not disrupt re-pairing of precopula pairs. The brief exposures to 1.0 (for 20 min) and 2.0 mg/L of lindane (2–20 min) resulted in significantly reduced feeding rates during the first 24 hour post-exposure, and re-pairing was affected by treatments that combined higher concentrations and longer exposures. The median survival time for briefly exposed animals varied with concentration and exposure period (Malbouisson et al. 1995).

The investigation of combinations of behavioral endpoints, such as locomotory and ventilatory activities, avoidance and drift away from predators, and/or pollutants and/or parasites, may be a very promising direction for future behavioral ecotoxicology studies. It is anticipated that such studies may provide a more realistic assessment of the consequences of pollution.

6 Mode-of-Action Studies and Biomarkers

Some recently published studies with gammarids and related species suggest that *Gammarus* spp. may not only be suitable for nonspecific chronic toxicity testing for pollution-induced impairment of feeding activity, behavior, or development and sometimes mortality, but may also be useful for the assessment of more subtle, mode-of-action-driven chronic toxicity patterns. Such patterns can be investigated by using specific, mode-of-action-related endpoints or biomarkers at several organization levels.

At the cellular level, pollutant exposure may inhibit energy production (oxidative phosphorylation) or enzyme activity, or cause gene toxicity, carcinogenic activity, and oxidative stress. Some cellular responses can be measured by using biomarkers. Examples of potential biomarkers are metallothionein (MT) that protects the organism against metal-induced toxicity, stress proteins (heat shock proteins) that protect cells, and acetylcholinesterase, Na^+/K^+-ATPase, and Ca^{2+}-ATPase that protect against certain types of neurotoxicity.

At the individual and population levels, nonspecific chronic effects, such as those mentioned above (behavior and feeding activity), may occur. Effects with specific known modes of action may also occur. By using biomarkers for a set of specific endpoints, it is possible to evaluate a pollutant for its potential to induce several effects, e.g., developmental toxicity, chronic toxicity, immunotoxicity, and endocrine disruption. Table 4 provides a list of methods used to assess bioenergetic responses and effects on reproduction at the population level, whereas Table 5 summarizes biomarkers available for use in detecting different specific endpoints, including endocrine disruption.

6.1 Bioenergetic Responses, Excretion Rate and Respiration Rate

6.1.1 O:N Ratio, Respiration, and Ammonia Excretion

Gammarids have been used to assess the impact of oil and oil dispersants on a model littoral ecosystem in the Baltic Sea (Carr and Linden 1984). Bioenergetic (O:N ratio) measurements were made for *G. salinus;* ammonia excretion and respiration rates were also measured. No effects on ammonia excretion rates, respiration rates, or O:N ratios were observed after 1 day of exposure. However, after 10 days, highly significant differences were recorded between experimental and control groups for all three parameters, showing that both oil and oil/dispersant treatments produced subtle physiological alterations. Interestingly, the use of a chemical dispersant apparently resulted in a more rapid recovery of *G. salinus* than would have occurred if the oil had not been chemically dispersed (Carr and Linden 1984). Olsen et al. (2008) measured cellular energy allocation (CEA) in the sea ice amphipod *G. wilkitzkii* after this organism was exposed for 1 month to the water-soluble fraction (WSF) of oil. With the CEA biomarker, one is able to measure the energy budget of organisms by biochemically assessing changes in carbohydrates, protein and lipid content, as well as changes in electron transport system activity. The authors observed a significantly higher protein content at the medium dose compared to controls, but no effects were observed on the total energy budget, indicating that parts of the energy budget of *G. wilkitzkii* were affected by a WSF component of oil (Olsen et al. 2008).

6.1.2 Energy Input and Output with "Scope-for-Growth" Assays

Another approach for assessing stress was introduced by Naylor et al. (1989). The "scope-for-growth" (SfG) approach (Bayne et al. 1979) uses the difference between the energy input to an organism from its food and the output from respiratory metabolism to provide a good physiological measure of stress; in principle, this approach is directly related to population and community processes (Naylor et al. 1989). The rationale behind this approach is that physiological processes can often be assessed more easily and precisely than population and community ones, and

Table 4 Test methods for mode-of-action-related endpoints with gammarids

Species	Test substance/media	Exposure type	Exposure period	Endpoints/biomarkers	Effect/effect concentration	Remarks	Reference
Bioenergetic responses, excretion rate, and respiration rate							
G. salinus immature males	Baltic sea water + crude oil 20 mg/L + crude oil 20 mg/L and oil dispersants (15.4 mL)	Continuous seawater flow into circular pools	12 days	Ammonia (NH4), excretion rate and respiration rate, O:N ratio	Crude oil +/- dispersant: significant NH$_4$ increase, significant O$_2$ increase, significant O:N decrease	The use of a chemical dispersant resulted in a more rapid recovery	Carr and Linden (1984)
G. wilkitzkii ovigerous females	Water-soluble fraction (WSF) of oil: high: PAH 55–8 ppm; medium: PAH 10–2 ppm; low: PAH 5–1 ppm	Continuous flow-through	1 month	Cellular energy allocation (CEA)	10–2 ppm PAH: significantly higher protein content	No effects on the total energy budget	Olsen et al. (2008)
G. pulex males	Zinc (Zn): 0.3, 0.5, 0.7 mg/L Low pH: pH 5	Scope of growth (SfG)	5 days	SfG: Food absorption (A) – energy output (respiration, R)	A and SfG significantly reduced by 0.5 and 0.7 mg/L Zn and pH 5	Most sensitive endpoint: A	Naylor et al. (1989)
G. pulex males	Zn: 0.3, 0.5, 0.7 mg/L 3,4-Dichloroaniline (DCA): 0.125, 0.25, 0.5, 1 mg/L Oxygen: 100, 75, 50% saturation Ammonia (NH$_3$): 0.07 mg/L	Scope of growth (SfG)	6 days	SfG: Food absorption (A) – energy output (respiration, R)	A and SfG significantly reduced for Zn (>0.5 mg/L), DCA (>0.5 mg/L), oxygen (>50%), ammonia (0.07 mg/L)	Most sensitive endpoint: A	Maltby and Naylor (1990)

Table 4 (continued)

Species	Test substance/media	Exposure type	Exposure period	Endpoints/biomarkers	Effect/effect concentration	Remarks	Reference
G. pulex males	In situ: control upstream vs. polluted downstream site	SfG – in situ	6 days	SfG: Food absorption (A) – energy output (respiration, R)	In situ: A and SfG significantly reduced at downstream-polluted site	Most sensitive endpoint: A	Maltby et al. (1990)
G. pulex brooding females	Zn: 0.1, 0.3, 0.5 mg/L	SfG and reproduction	7 days	SfG: Food absorption (A) – energy output (respiration, R). Reproduction	>0.3 mg/L Zn: A and SfG significantly reduced, significant decrease in the size of released offspring from subsequent brood, increased no. of aborted broods	No effect on the number or the size of released offspring from current brood	Maltby and Naylor (1990)
G. pulex	Copper (Cu): 11.2, 14.6, 18.2, 23.1 µg/L	Flow-through	100 days	Growth, population density (PD), age composition (AC), number of adults (NoA)	Increase in Cu, decrease in density and number of juveniles. LOEC: PD: 14.6 µg/L Cu, AC: 14.6 µg/L Cu, NoA: 18.2 µg/L Cu	Control and 11.0 µg/L Cu, the initial density doubled and mainly juveniles were present	Maund et al. (1992)

Table 4 (continued)

Species	Test substance/media	Exposure type	Exposure period	Endpoints/biomarkers	Effect/effect concentration	Remarks	Reference
G. pulex copulatory pairs and newly released juveniles	Esfenvalerate (ESV): 0.05, 0.1, 0.3 μg/L	Reproduction test	1 hour pulse + 2 weeks post-exposure	Survival, pairing behavior, reproductive output	0.1–0.6 μg/L ESV: decreased survival, pairing behavior, reproductive output. 0.05 μg/L ESV: immediate disrupted reproducing pairs, egg or offspring release from brood pouch, delays in pair formation and reproduction following transfer to clean water		Cold and Forbes (2004)
G. locusta juveniles (2–4 mm)	Cu-spiked sediments: 1.4, 2.8 and 4.9 mg/kg dry weight	Static renewal, seawater and spiked sediment	28 days	Cu body burden, metallothionein (MT), growth, reproduction	4.9 mg/kg Cu: stimulation of growth and reproduction, bioaccumulation of 95 μg Cu/g dry weight, significant synthesis of MT (1.7 mg/g dry weight) in males	Cu contamination seems to lead to an unexpected condition improvement	Correia et al. (2001)

Table 4 (continued)

Species	Test substance/media	Exposure type	Exposure period	Endpoints/biomarkers	Effect/effect concentration	Remarks	Reference
G. wilkitzkii females with embryos in brood pouch	Water-soluble fraction (WSF) of oil, PAH concentrations: high dose: 55–8 ppm; medium dose: 10–2 ppm; low dose: 5–1 ppm	Embryogenesis	30 days	Reproductive stage, eggs per female, weight, developmental malformations	High dose: frequency of embryo aberrations significantly higher compared to controls	No significant differences in reproductive stage, eggs per female or weight	Camus and Olsen (2008)
C. marinus freshly fertilized embryos	Cu, pentachlorophenol (PCP), benzo[a]pyrene (B[a]P)	In vitro embryo culture, aqueous exposure	Until embryos hatched or development ceased	Embryo development length, width, stage	Significantly impaired embryo development at 20 µg/L Cu, 20 µg/L PCP, 20 µg/L B[a]P Cu, PCP: extended embryogenesis by 4–8 days. B[a]P: hatched at the same time as controls but were significantly smaller	Stages 2–4 were all prolonged by Cu, PCP, B[a]P and the time to complete stage 5 was reduced	Lawrence and Poulter (2001)
G. fossarum	Sewage treatment plant (STP) discharging effluent containing xenoestrogens: locations: Lu, Ku above: control Ld, Kd below: exposure	Population structure investigation by field sampling	1.5 years, field sampling every 2 (Lu, Ld) and 4 (Ku, Kd) weeks	Sex ratio, intersexuality, population structure	No effect on sex ratio and intersexuality. Proportion of breeding female gammarids downstream (Kd, Ld) was tentatively lower than upstream (Ku, Lu).	Kd, Ld: tendency toward a decreased proportion of smallest juvenile gammarids in the population compared to Ku, Lu	Ladewig et al. (2006)

Table 4 (continued)

Species	Test substance/media	Exposure type	Exposure period	Endpoints/biomarkers	Effect/effect concentration	Remarks	Reference
G. pulex, A. aquaticus (2-week-old juveniles)	Landfill leachate [1200 mg/L chemical oxygen demand (COD) and 600 mg/L biological oxygen demand$_5$ (BOD$_f$)]	Reproduction experiment, static renewal	4 months	Reproduction	A dilution as high as 1:66 influences the fecundity of the Gammarus population. A dilution of 1:20 affects the size of the Asellus breeding colony		Bloor et al. (2005)

Table 5 Biomarkers for multiple stressors in *Gammarus* spp.

Abbreviation	Name	Species	Function	Toxicant: activity/effect	Reference
Molting process and exoskeleton integrity					
Chitin	Chitin, glycosaminoglycans	*Gammarus* sp.	Addition of chitin during the endocuticle production between the cuticle and the epiderm, ecdysone-mediated chitin formation process, or glucosamine formation	Xenoestrogens: ↑ chitin Polluted marina: ↑ chitin	Gagné et al. (2005)
Arthropodin	Neutral arthropodin	*Gammarus* sp.	Post-molt phase, arthropodin inclusion in the cuticle supports the exoskeleton-hardening process	Polluted marina: ↑ arthropodin	Gagné et al. (2005)
Sclerotin	Alkali-extractable sclerotin	*Gammarus* sp.	Post-molt phase, sclerotin inclusion in the cuticle to support exoskeleton-hardening	Polluted marina: ↑ sclerotin	Gagné et al. (2005)
ASP	Acid-soluble proteins	*Gammarus* sp.	Post-molt phase, ASP production assists calcareous layer formation	Polluted marina: ↑ ASP	Gagné et al. (2005)
ALP	Alkali-labile phosphates	*Gammarus* sp.	alkali-labile phosphates in proteins, important for phosphate mobilization	Polluted marina: ↑ ALP	Gagné et al. (2005)
Endocrine disruption					
Vtg	Vitellogenin-like proteins	*Gammarus* sp.	Egg-yolk protein precursor, oocyte-maturation stage	Polluted marina: ↑ Vtg	Gagné et al. (2005)
hsp90	Heat-shock protein 90	*G. fossarum*	Steroid receptor interactions and the modulation of sex hormone signal transduction level of stress protein reflects the maturity stage of the oocytes	Bisphenol-A: ↑hsp90	Schirling et al. (2006)

Table 5 (continued)

Abbreviation	Name	Species	Function	Toxicant: activity/effect	Reference
Energy demand/glucose metabolism/lipid anabolism					
G6PD	Glucose-6-phosphate dehydrogenase	*Gammarus* sp.	Intermediary metabolism of glucose, lipogenic metabolism enzyme	Polluted marina: ↑ G6PD + ↑ ME ↑ energy demand, lipid anabolism	Gagné et al. (2005)
ME	NADH-generating malic enzyme	*Gammarus* sp.	Intermediary metabolism of glucose, lipogenic metabolism enzyme	Polluted marina: ↑ G6PD + ↑ ME + ↑ Vtg: delay in gametogenesis	Gagné et al. (2005)
ICD	Isocitrate dehydrogenase	*Gammarus* sp.	Aerobic metabolism of glucose, lipogenic metabolism enzyme	Polluted marina: ↑ G6PD + ↑ ME + ↑ ICD Increased glucose metabolism	Gagné et al. (2005)
General stress response					
hsp70	Heat shock protein 70	*G. fossarum*	Sensitive biomarker induced by various proteotoxic stressors	Cd^{2+} and 3BC: ↑ hsp70, low + intermediate concentrations ↑ hsp70, high concentrations, pathological damage	Scheil et al. (2008) Schill et al. (2003)
Metal exposure/cellular stress					
MT	Metallothionein	*G. locusta*	Homeostasis and detoxication, indicator for cellular stress	Water-borne Cu: ↑ MT	Correia et al. (2002)
Oxidative stress					
LP	Lipid hydroperoxides	*G. locusta*	Oxidative damage: peroxidation of unsaturated lipids to cytotoxic lipic hydroperoxides from reactive oxidative species (ROS) (i.e., metals)	Water-borne Cu: ↑ LP 1–4 d), ↑ LP (4–6 d) Sediment Cu: ↑ LP, after 4 days of exposure	Correia et al. (2002)

Table 5 (continued)

Abbreviation	Name	Species	Function	Toxicant: activity/effect	Reference
Neurotoxicity/pesticide exposure					
ChE	Cholinesterase	*G. pulex*	ChE and AChE degrade the neurotransmitter acetylcholine in cholinergic synapses	Pirimiphos-methyl: ↑ ChE	McLoughlin et al. (2000)
AChE	Acetylcholinesterase	*G. pulex*		Fenitrothion: ↑ AChE	Streit and Kuhn (1994)
		G. pulex		Parathion: ↑ AChE	Crane et al. (1995)
		G. pulex		Malathion 60: ↑ AChE	Xuereb et al. (2007)
		G. pulex		Chlorpyrifos: ↑ ChE	Maltby and Hills (2008)
				Chlorpyrifos: ↑ ChE	
				Cypermethrin: ↑ ChE	
Detoxification					
GST	Glutathione-S-transferase	*G. pulex*	Detoxification enzyme	Lindane: ↑ GST Permethrin: ↑ GST	McLoughlin et al. (2000)

although toxic substances affect the physiological processes of individual organisms, their ecological impacts occur at the population and community levels. The SfG was measured in *G. pulex* organisms which were exposed to zinc (0.3, 0.5, and 0.7 mg/L) and low pH (pH 5); reported results were that both treatments significantly reduced the SfG of individuals, and the most sensitive component of the energy budget was food absorption (Naylor et al. 1989).

Another group also investigated the use of SfG for freshwater systems (Maltby and Naylor 1990). The authors exposed *G. pulex* to conditions, often associated with pollution, by assessing the effects of four specific substances: a metal (zinc), an organic compound (3,4-dichloroaniline), and two dissolved gases (oxygen and ammonia). In all cases, SfG was reduced by the stress, primarily from a depression of energy intake. The energy output (respiration) was significantly affected only by ammonia (Maltby and Naylor 1990). Maltby et al. (1990) then deployed the SfG system to examine whether it would be an equivalent sensitive indicator of stress for *G. pulex* in the field, as it had under laboratory conditions. In every case, SfG was reduced at downstream-polluted sites compared to upstream reference sites. This reduction in SfG was the result of a decrease in energy intake (absorption) rather than an increase in energy expenditure (respiration). Maltby and Naylor (1990) used *G. pulex* brooding females to compare the effect of Zn on SfG and on reproduction. At 0.3 mg/L of zinc (and higher), SfG was significantly reduced, which resulted from significant decreases in energy absorption. The offsprings released from the subsequent brood were smaller, but there was no effect on size or number of offspring released; this result was true for both current and subsequent broods. However, both present and past zinc stress caused an increase in the number of broods aborted (Maltby and Naylor 1990).

6.2 Population Experiments, and Development and Reproduction Modeling

6.2.1 Growth, Density, and Age Composition

Gammarids were used to investigate the chronic toxicity of metals, e.g., copper. *G. pulex* populations were exposed for 100 days to copper concentrations below the 240-hour LC_{50} for juveniles (Maund et al. 1992). Copper significantly affected growth, density, and age composition of the populations and the effects were dose dependent. In the control and the lowest treatment (11.0 μg/L) groups, the initial population density doubled and mainly juveniles were present. With increasing copper concentration, a decrease in both density and number of juveniles present was observed. The density was lower than that of the initial population. At the highest concentration (23.1 μg/L Cu), the number of adults was significantly reduced. The LOEC values for population density, age composition, and number of adults were 14.6, 14.6, and 18.2 μg/L Cu, respectively.

6.2.2 Life-History Traits

Different life stages of *G. pulex* were used to examine effects on key life-history traits following short and environmentally realistic pulse exposures of the pyrethroid insecticide esfenvalerate (Cold and Forbes 2004). Concentrations in the range of 0.1–0.6 µg/L for as little as 1 hour affected *G. pulex* survival, pairing behavior, and reproductive output that still could be detected at least 2 weeks following the pulse. Exposure to 0.05 µg/L for 1 hour induced immediate disruption of reproducing pairs, release of eggs or offspring from the brood pouch, and substantial delays in pair formation and subsequent reproduction, following transfer to clean water (Cold and Forbes 2004).

6.2.3 Population Endpoints Combined with Body-Burden and Metallothionein Induction

In another study, the effect of copper-spiked sediments on *G. locusta* was evaluated during 28 days of exposure (Correia et al. 2001); key measures included copper body-burden and metallothionein (MT) induction and an integration of these with organism and population-level endpoints. The most relevant sublethal effects detected in this study were stimulation of growth and reproduction at the highest treatment level (4.9 mg Cu/kg dry weight), bioaccumulation of Cu (95 µg Cu/g dry weight), and increased synthesis of MT (1.7 mg/g dry weight in males, $p < 0.001$), suggesting that the observed effects were associated with Cu contamination. The observed higher offspring production was regarded to be a direct consequence of faster growth rates. The authors suggest that hormesis is responsible for faster growth rates, which was an unexpected improvement induced by Cu contamination, because crustaceans need Cu in their blood pigment (hemocyanin; Gerhardt 1995, 1996).

6.2.4 Embryogenesis

Camus and Olsen (2008) studied malformations in embryos of the Arctic sea ice amphipod *G. wilkitzkii* exposed to the water-soluble fraction of oil. The female growth stages ranged from development stage three to nine, and no differences in reproductive stage were observed among the different treatments after 30 days of exposure. However, the frequency of embryo aberrations was significantly higher in the high-dose group compared to controls; this indicated that the embryos of *G. wilkitzkiii* were affected by exposure to the oil (Camus and Olsen 2008).

Embryogenesis was also used as an endpoint by Lawrence and Poulter (2001) to investigate the developmental toxicity of copper, PCP, and B[a]P toward the amphipod *C. marinus*. To conduct this study, the authors used *C. marinus* in an adjusted in vitro embryo culture method (Morritt and Spicer 1996). Maximal width and length of freshly fertilized and exposed embryos were measured daily and the developmental stage was determined for each individual until either development ceased due to disruption or the first juveniles hatched. Development of in vitro cultured embryos was significantly impaired by 20 µg Cu/L, 20 µg PCP/L, and 20 µg B[a]P/L. Cu and

PCP extended the period of embryogenesis by 4–8 days, whereas embryos cultured with B[a]P hatched at the same time as controls but were significantly smaller. Each pollutant affected specific stages, from Stage 2 onward. Stages 2–4, in which the embryo undergoes development of the germinal disc, dorsal organ rudiments, cordal furrows, appendage rudiments and segments, eye, and heart, were all prolonged in toxicant-exposed treatments. Generally, the time to complete Stage 5 was reduced in pollutant-exposed embryos. The results indicate that both swimming stamina and embryogenesis may be used in amphipods as sensitive bioassays for toxic effects (Lawrence and Poulter 2001).

6.2.5 Population Structure and Dynamics

Population structure and population dynamics of *G. fossarum* (Ladewig et al. 2006) were investigated in a field experiment. Gammarids were sampled at two streams in Germany, each with two sampling sites above and below a sewage treatment plant (STP) that discharged effluents known or assumed to have endocrine-disrupting potential. Changes in the sex ratio of *G. fossarum* or occurrence of intersexuality was not observed in either stream, but differences in the structure and dynamics of *G. fossarum* populations were found, and these were more pronounced in one stream. This result agrees with findings on *G. pulex* populations, which significantly differed in population density, standing crop biomass, individual size, and sex ratio in streams with different lotic conditions, suggesting that some of those dissimilarities were caused by pollutants (Crane 1994).

Interestingly, in a study conducted by Ford et al. (2006), with *E. marinus* populations from polluted and reference sites, consistently higher level of intersexuality was found throughout the year at sites receiving industrial contaminants when compared with reference sites. Infection of *E. marinus* by microsporidian parasites appeared to be more prevalent at impacted sites, indicating that parasitism may partly be responsible for intersexuality but may not be its only cause. Other environmental factors probably exist. Whether pollution can cause intersex directly remains to be confirmed, although, due to the apparent fragility of sex determination in the Amphipoda, it cannot be ruled out (Ford et al. 2006).

Possible negative effects on *G. pulex* and *A. aquaticus* populations were investigated to determine the potential ecological implications of leached contaminants reaching the water table (Bloor et al. 2005). A specific landfill leachate [1200 mg/L chemical oxygen demand (COD) and 600 mg/L 5-day biological oxygen demand ($BOD_{(5)}$)] was used to develop a standardized long-term sublethal ex situ toxicity-testing program with juveniles. The authors found that a dilution even as high as 1:66 influenced the fecundity of a *Gammarus* population, while a dilution of 1:20 affected the size of an *Asellus* breeding colony (Bloor et al. 2005).

6.2.6 Population Experiments Combined with Modeling on Reproductive Output

The author of a dissertation at the Technical University of Dresden (Germany) investigated the activity of environmental chemicals toward *G. fossarum,* focusing on population experiments and on an individual-based reproduction model (Schmidt 2003). Artificial indoor streams, containing *G. fossarum* and other aquatic invertebrates to simulate a community, were used to investigate the herbicide terbutryn and the insecticide fenoxycarb. No statistically significant effects on population-related parameters were found for these chemical, but at the highest exposure concentrations of each (289 μg/L terbutryn and 50 μg/L fenoxycarb), chronic toxic effects were reported, although they were not statistically significant. The nature of this chronic effect was decreased formation of precopula pairs and a reduced number of offspring. Levels of terbutryn in excess of 2 μg/L decreased growth in a dose-dependent manner, probably because of a dose-dependent decreasing algae availability, which comprised a food source. A reproduction model (GamMod) was also developed that incorporated the following fenoxycarb exposure parameters: the number of juvenile offspring/female (control, 10.5; 0.05 μg/L, 10.0; 0.5 μg/L, 7.6; 5 μg/L, 8.0; and 50 μg/L, 10.3), the duration of brood development (28–29 days), and juvenile mortality (control, 6 days; 0.05 μg/L, 12 days). By comparing the experimental and modeling data it was possible to show that the GamMod model gave a plausible description of the population dynamics of *G. fossarum*. GamMod was able to describe the measured data well, with a geometric performance index between 1.6 for controls and 2.7 for the highest exposure concentration. A sensitivity analysis showed that juvenile mortality is the most sensitive parameter in this experimental setup. By evaluating several different parameters (e.g., juvenile mortality, sex ratio, and number of juveniles/female), it was possible to show how the model could be useful for calculating outcomes of possible scenarios, such as reducing the number of females in a population.

6.3 Endpoints and Biomarkers for Endocrine Disruption in Gammarids

Until the present, investigations involving endocrine-disrupting compounds (EDCs) in aquatic ecosystems have generally been performed on fish and mollusks. Some research, however, has been dedicated to other organisms, including arthropods. During certain periods of somatic growth, arthropods periodically go through molting (ecdysis). During such periods of growth, gametogenesis, production of a new cuticle, and shedding of the old exoskeleton are physiologically regulated and are especially vulnerable to endocrine disruption (Schirling et al. 2004). For example, during the growth phase of adult female crustaceans, hormonally controlled vitellogenesis occurs after ecdysis (Subramoniam 2000). Molting is controlled by a steroid hormone 20-hydroxyecdysone (Baldaia et al. 1984), but its involvement in vitellogenesis is controversial. During oocyte growth, various arthropods were found to have elevated levels of vertebrate-type steroids such as progesterone, testosterone, and 17β-estradiol (Cardoso et al. 1997; Fairs et al. 1990). However, treating *D. magna* with the estrogenic diethyl phthalate inhibited molting but did not disrupt vitellogenin (Vtg). This indicates that estrogenic vertebrate EDCs also interfere with the arthropod endocrine system by acting through the ecdysteroid receptor (Zou and Fingerman 1997). To complicate the issue further, ecdysone not only is important for molting but also plays a major role during embryogenesis, where it is bound to Vtg in oocytes (Subramoniam et al. 1999). Chitin synthesis, which plays an important role during molting, was also found to be controlled by the ecdysone receptor pathway (Gagou et al. 2002; Nakagawa et al. 1995).

Mechanistic studies on the interaction of potential xenohormones with the endocrine system are still lacking for gammarids. There are only a limited number of publications for EDC in crustaceans, but those provide early ideas on what may become suitable endpoints, biomarkers, exposure scenarios, and ways to assess endocrine disruption in this genus.

6.3.1 Vitellogenin-Like Proteins and Lipogenic Enzymes (ICD, ME, G6PD)

Vtg, the energy-rich egg-yolk protein precursor, has been proposed as a biomarker to characterize the maturation stage of oocytes in crustaceans (Chang and Jeng 1995; Oberdörster et al. 2000). In addition, it was proposed that the increase in energy demand (glucose metabolism) leading to lipid anabolism during gametogenesis could also be followed by measuring intermediary lipogenic metabolism enzymes such as glucose-6-phosphate dehydrogenase (G6PD), isocitrate dehydrogenase (ICD), and the NADH-generating malic enzyme (ME) (Mori 1967; Sunny et al. 2002). Gagné et al. (2005) investigated the suitability of those biomarkers to assess changes in gametogenesis by collecting gammarids at polluted sites. They found that females from polluted sites had increased levels of Vtg-like proteins, indicating a delay in gonad maturation from the presence of environmental contaminants. In addition, increased lipogenic enzyme activities (ME and G6PD) were observed in females, supporting the delay-in-gamete-maturation hypothesis (Gagné et al. 2005). However, increases in aerobic metabolism of glucose (ICD) and intermediary metabolism of glucose (G6PD and ME) indicated an increased glucose metabolism, which is often observed in organisms exposed to pollutants (De Coen and Janssen 2003).

6.3.2 Heat Shock Proteins (hsp90) as Biomarkers for Endocrine Disruption in Gammarids

In vertebrates, it is known that the heat shock protein hsp90 is of crucial importance for steroid receptor interactions and modulation of sex hormone signal transduction (Pratt and Toft 1997). For invertebrates, the understanding of an equally complex system is still limited. The discovery of estrogen receptors in invertebrate taxa (De Waal et al. 1982) indicates that steroid-binding proteins and, therefore, mechanisms associated with the signal transduction process are phylogenetically very old (Thornton et al. 2003). Schirling et al. (2004) selected hsp90 as a potential biomarker for endocrine disruption, although it is well known that hsp90, like all stress proteins, also responds to stressors that do not target the endocrine system. To control for this, the authors also measured the well-established general stress marker hsp70, as a "nonspecific stress effect control" for hsp90. For 12 weeks, animals were removed every 14 days to histologically determine the stage of the reproductive cycle and to measure variations in levels of hsp70 and hsp90 by using an immunoblotting assay. The maturation stage of the female gonad was identified from the structure of the oocytes, confirming that an almost complete reproductive cycle occurred within 12 weeks. Hsp70 and hsp90 levels were found to be inversely correlated over the course of the reproductive cycle. The authors reported that the hsp90 level at the beginning of the reproductive cycle was low, whereas the hsp70 level was at its peak. At the end of the cycle, when mature oocytes were present, the opposite was true. The finding that levels of stress proteins reflect the maturity stage of the oocytes provides prerequisite baseline information that may become quite useful. Such information enhances the ability to interpret biomarker studies on endocrine effects of chemicals in gammarids.

6.3.3 Chitin as a Biomarker for EDC Effects on Molting

Parts of the molting process, e.g., chitin synthesis and characteristics of exoskeleton protein and carbohydrate, are at least partially controlled by the molting steroid hormone ecdysone (Gagné and Blaise 2002; Gagou et al. 2002; Nakagawa et al. 1995). Indeed, incorporation of glucosamine into the integument was shown to be enhanced by ecdysone (Nakagawa et al. 1995). During the post-molt phase, the cuticle precursor is mainly composed of a flexible and transparent organic matrix and is usually devoid of chitin (glycosaminoglycans; Nation 2002). To support the exoskeleton-hardening process, neutral (arthropodin) and alkali-extractable (sclerotin) proteins are incorporated into the cuticle. At the inter-molt and post-molt stages, the endocuticle, between the cuticle and the epiderm, is produced, wherein the addition of chitin, sulfur, and calcium phosphate takes place (Nation 2002; Vigh and Dendinger 1982). During crustacean ecdysis, the cuticle is gradually dissolved so that the relative proportion of chitin increases as salt and protein content decreases. Thus, Gagné and Blaise (2002) propose that the integrity of ecdysis and the maturation state of exoskeletons could be followed by measuring the relative proportions of the acid-extractable proteins sclerotin, arthropodin, and chitin in exoskeletons, and phosphate level in proteins of the arthropod epidermis (Gagné and Blaise 2002).

The notion that xenoestrogens can affect molting in invertebrates was supported in fiddler crabs, where synthetic estrogens (i.e., diethyl phthalate, 4-(*tert*)-octylphenol, and 2,4,5-trichlorobiphenyl) were found to reduce molting-relevant chitobiose activity in the epidermis and hepatopancreas, and increase the proportion of chitin in the exoskeleton (Zou and Fingerman 1999).

A contrary result was observed by Gagné et al. (2005) when they collected *Gammarus* sp. individuals at four intertidal sites subjected to direct sources of pollution (marinas, ferry traffic, and harbors) and at one site with no direct source of pollution. Gammarids from polluted sites had significantly less chitin in exoskeletons, suggesting disruption in the ecdysone-mediated chitin formation process or decreased formation of glucosamine from glucose (Gagné et al. 2005).

6.3.4 Sex Ratio and Precopula Pairs

Watts et al. (2002) investigated how an artificial estrogen 17α-ethinyl estradiol (EE2) affects the sex ratio in cultures of freshwater crustaceans of *G. pulex*. Mixed populations of 90 individuals were exposed to 0.1, 1, and 10 µg/L EE2 for 100 days in a flow-through system. In all treatment groups, population size dramatically increased due to recruitment, with neonate and juvenile gammarids being the most abundant. Mean population sizes in the solvent control (257) and 0.1 µg/L EE2 treatment groups (267) did not differ from standard controls, but at the 1 and 10 µg/L EE2 exposures groups (385 and 411, respectively), population numbers were significantly greater than in the control population (169). The sex ratio of adults for all EE2 treatments was greater than 2:1 (female:male), with significantly more females than in the controls. The number of male adults, precopula guarding pairs, and ovigerous females did not differ among treatments. Secondary antennal and gnathopod length in males was consistently greater than in females, but no other differences were found between groups (Watts et al. 2002).

Effects on sex ratio were also detected in *G. pulex* exposed to estrogenic substances released by a sewage treatment plant, where an increased proportion of females was observed, along with an abnormal oocyte structure during vitellogenesis (Gross et al. 2001).

In another study, Watts et al. (2001) investigated the effects of EE2 and bisphenol-A (BPA) on survival and reproductive behavior of *G. pulex*. Reproductive behavior, like ability of males and females to detect each other to form precopulatory guarding pairs and to continue guarding behavior, was disrupted only at relatively high concentrations (3.7 mg/L EE2 and 8.4 mg/L BPA). This was probably caused by general toxicity rather than an endocrine-mediated process, indicating that precopulatory guarding in acute exposures is not a suitable endpoint for detecting EDCs (Watts et al. 2001).

6.3.5 Gonad Histology and hsp90

BPA was investigated by Schirling et al. (2006) regarding how it affects stress protein levels (hsp70 and hsp90) and gonad histology of *G. fossarum* in artificial indoor streams. Exposure to 50 and 500 µg/L of BPA resulted in accelerated maturation of oocytes in females and in a decline in the number and size of early vitellogenic oocytes. The level of hsp90, which plays a pivotal role in vertebrate sex steroid signal transduction, was significantly reduced by BPA at those concentrations (Schirling et al. 2006). This result is in line with the strong co-variation of hsp90 level with the reproductive cycle (Schirling et al. 2004). The authors reported that in early stages of oocyte development, the hsp90 level of individuals was three times lower than in specimens with mature eggs. BPA seems to disrupt this correlation and leads to reduced hsp90 levels and accelerated maturation of oocytes.

In a field study, again using gonad histology and hsp90, individuals of autochthonous populations of *G. fossarum* were examined for their maturity status, oocyte development, and biochemical parameters associated with their reproductive cycle (Schirling et al. 2005). Despite the isolated investigation of different size classes of *G. fossarum* individuals to reduce data variability, a high variability was recorded, which prevented observation of significant differences for most of the measured parameters. Nevertheless, effects on gonad development and hsp90 level, both parameters related to the endocrine system, were found in the Kçrsch river. Downstream from the discharge of treated sewage, larger late vitellogenic oocytes, increasing atresia, and decreasing hsp90 levels were observed. This corresponds well with a higher estrogenic potential introduced by those effluents (Jungmann et al. 2004), compared to the other site, Lockwitzbach, wherein no effects were found.

6.3.6 Gametogenesis Activity and Gonad Maturation

In yet another field study, *Gammarus* sp. individuals were collected at four intertidal sites that were subjected to direct sources of pollution and at one site with no direct source of pollution. Subsequently, levels of Vtg-like proteins, MT, alkali-labile phosphates (ALPs) in proteins, and lipogenic enzyme activities (i.e., glucose-6-dehydrogenase, isocitrate dehydrogenase, and malate enzyme) were measured in whole soft tissues (Gagné et al. 2005). In addition, levels of pH-dependent extractable protein and chitin were determined in the exoskeleton to assess potential impacts of pollution on exoskeleton integrity and the molting process. The authors found that whole-body weights of both sexes were significantly lower at polluted sites and that females displayed either induced or decreased Vtg-like proteins at polluted sites, indicating significant changes in gametogenesis activity (Gagné et al. 2005). MT levels were not sex dependent and tended to be induced at all affected sites. At some impacted sites, females had a tendency toward higher ALP levels, indicating altered phosphate mobilization at those sites. In addition, lipogenic enzyme activities were increased at impacted sites for both sexes, suggesting a delay in gonad maturation rates. Principal component analysis revealed that gammarids collected at affected sites displayed substantial changes in the proportion of chitin, arthropodin, sclerotin, MTs, and intermediary glucose metabolism (glucose-6-phosphate dehydrogenase and isocitrate dehydrogenase in soft tissues) and thus suffered from disturbed gametogenesis and exoskeleton integrity (Gagné et al. 2005).

6.4 Other Specific Biomarkers for Detecting Multiple Stressors in Gammarids

6.4.1 Heat Shock Proteins (hsc/hsp70) as Biomarkers for Stress Response in Gammarids

Cells from virtually all organisms respond to a variety of stresses by the rapid synthesis of a highly conserved set of polypeptides called heat shock proteins (hsp), suggesting that hsp play an important role in both normal cellular homeostasis and stress response (Kregel 2002). Production of high hsp levels can be triggered by exposure to different kinds of environmental stress conditions, such as inflammation, exposure of the cell to toxins like trace metals, starvation, or hypoxia (oxygen deprivation). In most organisms, hsp70 is among the most prominent proteins induced as a stress response (Li and Werb 1982). Hsp70 is known to be a sensitive biomarker induced by various proteotoxic stressors (Nadeau et al. 2001; Triebskorn et al. 2002) and there has been a growing interest in tracing contaminant effects by tracking hsp70 levels in crustaceans (Arts et al. 2004; Köhler et al. 2000) and amphipods (Schill et al. 2003).

Scheil et al. (2008) examined the responses of hsp70 and of hepatopancreatic cells and cells of gut appendices in the freshwater amphipod *G. fossarum* following a short-term exposure (4 days) to five different concentrations of the chemical UV filter 3-benzylidene camphor (3-BC). Male and female gammarids showed increased hsp70 levels after exposure to low concentrations of 3-BC (0.033, 0.33, and 3.3 µg/L), with a maximum response at 3.3 µg/L, indicating physiological stress. Higher concentrations (33 and 330 µg/L) resulted in lower hsp70 levels, indicating an overwhelming stress response. This effect probably resulted from a cessation of hsp70 synthesis following pathological impact, as indicated by strong cellular responses and cellular damage obtained in epithelia of the hepatopancreas and the gut appendices after treatment with 330 µg/L 3-BC (Scheil et al. 2008).

By simulating a mining accident, Schill et al. (2003) investigated immediate stress responses to toxicants, by measuring two forms of the 70-kDa hsp hsc/hsp70 and assessing the recovery time an organism needs after the end of the exposure. During a 20-day experiment, adult *G. fossarum*, separated by sex, were exposed to nine cadmium concentrations for 5 days to simulate a short-term pulse of xenobiotics, followed by a recovery period of 15 days. Females were much more sensitive to cadmium than were males, and 4.28 ± 2.45 µg Cd^{2+}/L resulted in strong effects on survival rate of females, but not males. At the cellular level, cadmium induced an hsc/hsp70 response, with the lower Cd^{2+} concentrations leading to an induction of stress proteins, whereas higher Cd^{2+} concentrations resulted in a proportionately reduced hsc/hsp70 response, probably from pathological damage. Surviving individuals retained their capacity to induce stress protein production in the recovery period, even if the stress protein response system was overwhelmed by cadmium during the exposure period (Schill et al. 2003).

6.4.2 Metallothioneins and Lipid Peroxidation as a Biomarker for Metal Exposure and Oxidative Stress

Inducible MT has a key role in maintaining homeostasis and in detoxifying certain substances in aquatic invertebrates and other animals (Schlenk et al. 2000). MT is not only a very sensitive biomarker of exposure to certain metals but also an indicator of cellular stress through its role as a scavenger of organic free radicals and reactive oxygen species (ROS; Viarengo et al. 2000). There have been recent advances in facilitated research on MT in crustacean amphipods (Correia et al. 2001; Correia et al. 2002; Gagné et al. 2005). For example, Cu is needed for normal metabolic function but may become toxic if intracellular concentrations exceed the organism's requirements or detoxication capacity (Dos Santos Carvalho et al. 2004; Viarengo et al. 2000). The toxicity caused by Cu and other metals may derive from multiple mechanisms, including generation of ROS and induction of various types of oxidative damage, such as peroxidation of unsaturated lipids to cytotoxic lipid hydroperoxides (lipid peroxidation, LP)(Livingstone 2001).

Correira et al. (2002) exposed the marine amphipod *G. locusta* to sublethal concentrations of copper in water (4 days of exposure to 3, 5, and 10 mg/L Cu) or sediment (28 days of exposure to 1, 3, and 6 mg/kg Cu dry weight) and investigated the effects on putative MT and LP levels. In addition, they carried out a time-course exposure study (over 10 days) to a single water-borne concentration of Cu (4 mg/L). MT and LP were quantified by differential pulse polarography and as thiobarbituric acid-reactive malondialdehyde equivalents, respectively (Correia et al. 2002). MT was significantly induced by all water-borne Cu concentrations, but no increase in LP was observed in these animals. In contrast, LP levels increased in the time-course experiment within 1 day of exposure, peaked at 4 days, and returned to control value levels by day 6. Paralleling the decrease in LP, higher levels of MT were observed at days 6 and 10. In *G. locusta* exposed to Cu-contaminated sediments, no increase in MT levels was recorded, but significantly higher levels of LP were seen compared with controls. The observed inverse relationship between putative MT induction and the occurrence of LP indicates that MT may protect against the pro-oxidant effects of Cu. MT and LP may be suitable biomarkers for metal exposure and oxidative stress in gammarids.

6.4.3 Biomarkers to Assess Exoskeleton Integrity and the Molting Process

Exoskeleton integrity and molting can be assessed by measuring the levels of arthropodin, sclerotin, acid-extractable proteins, and chitin in high molecular weight proteins from whole tissues (Gagné et al. 2005). *Gammarus* sp. individuals were collected at four intertidal sites subjected to direct sources of pollution (marinas, ferry traffic, and harbors) and at one site with no direct source of pollution. The levels of pH-dependent extractable protein and chitin in the exoskeletons were used to assess the possible impacts of pollution on exoskeleton integrity and the molting process. Gammarids from contaminated sites had significantly higher levels of extractable proteins in their exoskeletons (i.e., arthropodin, sclerotin, and acid-soluble proteins) and a lower proportion of chitin in their exoskeletons at most impacted sites. It is possible that a disruption of chitin and pH-dependent protein mobilization led to disturbed exoskeleton integrity.

6.4.4 (Acetyl)-Cholinesterase Activity as a Biomarker for Neurotoxicity

Recently, the suitability of acetylcholinesterase (AChE) (Crane et al. 1995) and cholinesterase (ChE) (McLoughlin et al. 2000; Xuereb et al. 2007) activities as biomarkers of neurotoxic stress was investigated in the freshwater amphipod *G. pulex*. For decades, the inhibition of AChE and ChE activities has widely been used as biomarkers for the presence of certain pesticides (organophosphorus and carbamate) in aquatic species (Fulton and Key 2001). Such pesticides elicit their toxicity by inhibiting the activity of AChE or ChE enzymes, which are necessary to degrade the neurotransmitter acetylcholine in cholinergic synapses. This enzyme inhibition leads to accumulation of acetylcholine and thereby interferes with nerve function, inducing deleterious effects and eventually respiratory failure and death (WHO 1986). ChE activities have long been investigated in fish (Fulton and Key 2001). However, in aquatic invertebrates, classification of ChE isoforms has been investigated much less frequently than in vertebrates (Forget et al. 2002; Garcia-de la Parra et al. 2006; Varo et al. 2002). Moreover, several studies have pointed out the difficulty of using the ChE vertebrate classification scheme (AChE and BChE) for invertebrates, since invertebrate ChE retains characteristics of vertebrate forms and cannot be clearly distinguished from it (Varo et al. 2002).

McLoughlin et al. (2000) evaluated the usefulness of ChE as a biomarker in combination with feeding inhibition and mortality in *G. pulex* after exposing this species to zinc, linear alkylbenzene sulfonate (LAS; surfactant), lindane (organochlorine insecticide), pirimiphos-methyl (organophosphorus insecticide), and permethrin (pyrethroid insecticide). Lethality was the least sensitive endpoint. ChE inhibition was found to be a specific indicator of organophosphate exposure, although it was a considerably less sensitive biomarker (than 13-fold lower) than was feeding rate (McLoughlin et al. 2000). Exposure to 1.92 µg/L of the organophosphate pirimiphos-methyl led to a significant inhibition of enzyme activity after 24 hour, and the same was observed for 0.077 µg/L after a 48-hour exposure. Significant reductions in AChE activity have also been observed after exposing *G. pulex* to 1 µg/L each of the insecticides fenitrothion, parathion, and malathion for 24 hour (Crane et al. 1995; Streit and Kuhn 1994). However, no clear effects on *G. pulex* feeding rates or significant detrimental effects could be observed for the insecticide malathion (Crane et al. 1995). The inhibition of AChE activity by fenitrothion and parathion indicates that tolerance in various *Gammarus* species toward organophosphorus insecticides differs widely; the introduced species *G. tigrinus* showed a higher tolerance compared to the autochthonous species *G. pulex* and *G. fossarum*. This may help to explain recent changes in species composition (Streit and Kuhn 1994).

Xuereb et al. (2007) recently characterized ChE activity in *G. pulex* by using different substrates (acetylthiocholine iodide, propionylthiocholine iodide, and butyrylthiocholine iodide) and selective inhibitors (eserine sulfate, BW284c51, and *iso*-OMPA). The effect of chlorpyrifos, the widely used organophosphorus insecticide, on ChE activity was investigated. The results suggest that *G. pulex* possess only one ChE, which displays the typical properties of an acetylcholinesterase: it hydrolyses the substrate acetylthiocholine at a higher rate than all other tested substrates, and it is highly sensitive to eserine sulfate and BW284c51, but not to *iso*-OMPA (Xuereb et al. 2007). When *G. pulex* was exposed to realistic environmental concentrations of chlorpyrifos, significant AChE inhibition was observed; lethal effects appeared at inhibitions higher than 50%.

To understand impoverished stream communities in agricultural landscapes, Maltby and Hills (2008) used an experimental approach to investigate the effects of the insecticides cypermethrin and chlorpyrifos, which are possible candidates to contribute to such impoverishments. In this study, *G. pulex* were deployed during the application of the pesticides to the stream edge. The pesticides inhibited ChE enzyme activity, depressed feeding rate, and reduced survival. The authors found no clear insecticide-related effects on macroinvertebrate community structure or on the population densities of individual species. However, the adaptation of a no-spray buffer zone mitigated the individual-level effects (Maltby and Hills 2008).

The results of the above studies show the value of *G. pulex* as a sentinel organism for environmental assessment of sublethal neurotoxicity.

6.4.5 Glutathione-*S*-Transferase Activity as a Biomarker for Detoxification

The glutathione-*S*-transferases (GSTs) represent a major group of detoxification enzymes, and all eukaryotic species possess multiple cytosolic and membrane-bound GST isoenzymes (Hayes and Pulford 1995). Exposure to organochlorine compounds, such as the insecticides aldrin, endosulfan, or lindane, led to the induction of GST (Hans et al. 1993); other inducers are the pyrethroid insecticides such as cypermethrin (Gowland et al. 2002).

McLoughlin et al. (2000) evaluated the usefulness of the GST biomarker response in *G. pulex* after exposing this species to zinc, linear alkylbenzene sulfonate, lindane, pirimiphos-methyl, and permethrin. In addition, the authors evaluated feeding inhibition and mortality. A significant increase in GST enzyme activity occurred after 48 hour for both lindane (6.14 mg/L) and permethrin (0.12 mg/L). Thus, the GST biomarker performed with greater sensitivity but lower specificity, when compared with the ChE biomarker that was investigated in the same study. However, the more sensitive feeding rate was only marginally outperformed by the GST biomarker. The GST biomarker in *G. pulex* may be used as a rapid and sensitive indicator for toxicant exposure, but it has limited use as a diagnostic tool and provides only limited improvement in sensitivity over more ecologically relevant sublethal endpoints (e.g., feeding rate and growth rate) (McLoughlin et al. 2000).

When investigating impoverished stream communities in agricultural landscapes, Maltby and Hills (2008) evaluated how GST activity was affected by exposing *G. pulex* to the insecticides cypermethrin and chlorpyrifos. Contrary to the results on ChE inhibition, there was no significant difference on the GST activity in gammarids exposed to a combination of cypermethrin, isoproturon, and simazine (Maltby and Hills 2008). The reason for this may be that GST induction is less specific (responding to pyrethroid and organochlorine insecticides) and appears to have a longer response time (48 hour of permethrin exposure are needed for a significant induction; McLoughlin et al. 2000). The slow response time may explain why cypermethrin did not induce GST.

6.4.6 ATP Content as a Biomarker for Mycelium Species Composition on Gammarid Diet

The composition of the assemblage of fungi colonizing leaf material may be important in determining its quality as food for shredders like *G. pulex* (Barlocher and Kendrick 1975; Rossi 1985). Determining the species composition of mycelium patches on leaf material may enhance the understanding of fungus–invertebrate interactions, which are crucial to detritus processing in many freshwater bodies (Arsuffi and Suberkropp 1989; Maltby 1992). Bermingham et al. (1995) developed a monoclonal antibody-based (MAb) immunoassay for the detection and quantification of *A. longissima* that colonize leaf material, which allows for the determination of species-specific mycelium colonization. By using a co-immunization program, MAbs (to *A. longissima*) were raised in mice, then a cell line that produces a MAb of the immunoglobulin M class was cultured. This MAb was specific for *A. longissima*, both in an enzyme-linked immunosorbent assay (ELISA) and by immunofluorescence, but the immunoassay did not recognize other members of the aquatic hyphomycetes (Bermingham et al. 1995). This MAb (AL-HH8c) was then used to develop a quantitative ELISA in vitro. The antigen recognized by AL-HH8c is produced throughout the mycelium, irrespective of mycelial age and culture conditions. By using this MAb, mycelium of *A. longissima* colonizing leaf material can be detected (Bermingham et al. 1995).

7 Exposure Types

Our literature search showed that gammarids are used in different exposure scenarios. Often, similar experimental setups were used in the lab and in situ, with pollutants exposed via the aqueous, dietary, or sediment routes, either continuously or in pulsed exposure regimes (see Table 6).

7.1 Pulsed Exposure Assays and Models

Pulsed exposures of toxicants are often used in ecotoxicological studies because of their ability to describe the dynamics of toxicant exposure, as aquatic organisms experience them, in more detail. In addition, pulsed exposure has the advantage, depending on the test design, to show acute (during the pulse) as well as delayed (during a recovery phase between the pulses and at the end of the exposure) toxicity patterns.

7.1.1 Pulsed Exposure, Uptake, and Elimination of Pesticides in Lab and In Situ

Gammarids are often used in pulsed exposure experiments and seem to be useful organisms to determine uptake and elimination rates, and bioconcentration factors. Such studies were undertaken for chlorpyrifos and PCP by measuring internal concentrations of the two pesticides in *G. pulex* over a 3-day exposure phase and a subsequent 3-day elimination phase (Ashauer et al. 2006). Rate constants were obtained by fitting measured internal concentrations to a one-compartment, single first-order model. The uptake rate constants were 747±61 L/kg/day for chlorpyrifos and 89±7 L/kg/day for PCP. The elimination rate constants were 0.45±0.05 L/kg/day for chlorpyrifos and 1.76±0.14 L/kg/day for PCP. The resulting bioconcentration factors at steady state were 1660 and 51 for chlorpyrifos and PCP, respectively.

Table 6 Different exposure modes used to assess toxicant effects in gammarids

Species	Test substance/ media	Exposure type	Exposure period	Endpoints/ biomarkers	Effect/effect concentration	Remarks	References
Pulsed exposure assays							
G. pulex males and females	Chlorpyrifos (CPF), pentachlorophenol (PCP)	Pulsed exposure, static renewal	3-day exposure and 3-day elimination phase	Uptake rate, elimination rate, bioconcentration factor	*Uptake rate constants*: CPF: 747±61 L/kg/day PCP: 89±7 L/kg/day *Elimination rate constants*: CPF: 0.45±0.05 L/kg/day PCP: 1.76±0.14 L/kg/day *Bioconcentration factors*: CPF: 1660 PCP: 51	Uptake and elimination rate constants were estimated using ModelMaker	Ashauer et al. (2006)
Sediment toxicity assays							
G. locusta juveniles (2–4 mm)	Estuarine sediment toxicity Control, Muddy: T, P, D, sandy: S1, S2	Static renewal with seawater and sediment samples	28 days	Growth, reproduction, metallothionein (MT), DNA strand breakage	T, P: higher growth rates, improved reproductive traits. D, S1: DNA strand breakage, D: MT induction, S2: loss of DNA integrity, enhanced growth	S1 acutely toxic at 50% dilution, stimulated growth at 75% dilution	Costa et al. (2005) Neuparth et al. (2005)

Table 6 (continued)

Species	Test substance/media	Exposure type	Exposure period	Endpoints/biomarkers	Effect/effect concentration	Remarks	References
G. pulex, A. aquaticus (4–7 mm)	PAH-spiked sediment (fluoranthene, pyrene, chrysene, benzofluoranthene) with 30 mg PAH/kg dry weight	Choice experiment. Spiked vs. clean sediment	72 hour	Habitat choice	G. pulex and A. aquaticus avoided PAH-spiked sediment, the origin of the population (clean reference site or polluted site) did not affect habitat choice	Animals move away from the most polluted spots.	De Lange et al. (2006b)
In situ tests							
G. pulex, A. aquaticus	20 natural streams with differences in pH (4.3–7.5) and humic substances (color range: 8–280 mg Pt/L).	In situ with caged animals	25 days	Animal interactions on survival and physiological status	G. pulex: pH < 6.0: increased mortality. lower physiological status A. aquaticus: physiological status correlated with pH and significantly affected by humus	Under optimal conditions of high pH and low humus concentrations, species interactions seem asymmetric, where Gammarus decreases the survival and physiological status of Asellus	Hargeby (1993)

Table 6 (continued)

Species	Test substance/media	Exposure type	Exposure period	Endpoints/biomarkers	Effect/effect concentration	Remarks	References
G. pulex adult males	24 reference sites, 15 contaminated sites downstream of point-source discharges	In situ water quality biomonitoring with caged G. pulex	6 days	In situ feeding rate, ability of the assay to detect impacts of point-source discharges	Feeding rate inhibition between 27 and 99.6% downstream of point-source discharges	Reference sites: feeding rate strongly influenced by water temperature (76% of the variation) with a 30% feeding inhibition in summer (>90% power)	Maltby et al. (2002)

7.1.2 Pulsed Exposure Models

Ashauer et al. (2006) reviewed, evaluated, and compared two models that may be used to simulate effects, including survival, resulting from pulsed or fluctuating exposures of nontarget organisms to pesticides. The threshold damage model (TDM) and the time-weighted average (TWA) model were capable of simulating the observed survival of *G. pulex* (mean errors 15% or less, r^2 between 0.77 and 0.96) when exposed to two pesticides having contrasting modes of action (PCP and chlorpyrifos). The TDM proved to be particularly useful, because its parameters can be used to calculate recovery times, separate toxicokinetics from toxicodynamics, and parameter values reflect the mode of action (Ashauer et al. 2007b). When predicting the outcome of an exposure of *G. pulex* to carbaryl and using all exposure data (carbaryl, PCP, and chlorpyrifos) for fitting the models, the TDM outperformed the TWA model by facilitating an understanding of the underlying ecotoxicological processes and permitting calculation of recovery times (3, 15, and 25 days for PCP, carbaryl, and chlorpyrifos, respectively). In addition, the TDM enabled the prediction of long-term exposure effects after sequential pulses or fluctuating concentrations. This was also the case for the predicted outcomes of sequential pulsed exposure to carbaryl and chlorpyrifos (Ashauer et al. 2007a). The TDM predicted that recovery of damage to *G. pulex* from exposure to chlorpyrifos takes longer than that from exposure to carbaryl; therefore, the sequence of exposure matters and provides a process-based ecotoxicological explanation for the observed effects (Ashauer et al. 2007a).

7.2 Sediment Toxicity Assays

Gammarids and other benthic macroinvertebrates are often used in sediment toxicity assessment. The use of this species is detailed in guidelines issued by the American Society for Testing and Materials (ASTM 1993) and by SETAC in Europe (Hill et al. 1993). Gammarids are effective and sensitive indicators of ecosystem pollution (Munawar et al. 1989) and a substantial database on the responses of these macroinvertebrates to xenobiotics, nutrients, and other physicochemical perturbations exists (Burton et al. 1992). During their life cycle, gammarids spend extended periods of time in close contact with bottom sediments and in the water column above such sediments, which makes them susceptible to adverse effects lurking in contaminated sediments (Burton et al. 1992).

7.2.1 Sediment Tests with Marine Amphipods

A study was performed to evaluate the performance of *G. locusta* in chronic sediment toxicity tests (Costa et al. 2005). A multilevel assessment of chronic toxicity of estuarine sediments was used to integrate organismal- and population-level endpoints with biochemical marker responses. *G. locusta* were exposed for 28 days to five moderately contaminated sediments designated as follows: muddy: T, P, D;

sandy: S1, S2. The endpoints measured were survival, individual growth, and reproductive traits. Two of the muddy sediments (T, P) induced higher growth rates and improved reproductive traits, possibly because of the amount of organic matter in the sediment, which was nutritionally beneficial to the amphipods, while concurrently decreasing contaminant bioavailability. Biomarker responses did not reveal toxicant-induced stress in amphipods exposed to these sediments. One of the sandy sediments (S1) was acutely toxic at 50% dilution but stimulated amphipod growth at a dilution of 75% (Costa et al. 2005). The two sediments (D, S1) showed pronounced chronic toxicity, which affected survival and reproduction (female-based sex ratio and severely impaired offspring production). The biological marker responses of *G. locusta* showed the following (Neuparth et al. 2005): Two of the muddy sediments (T, P) did not cause chronic toxicity and were consistent with higher growth rates and improved reproduction at the population and organism level. Two other sediments (D, S1) exhibited pronounced chronic toxicity, affected DNA strand breakage, metallothionein induction (D), survival, and reproduction. The last sandy sediment (S2) exhibited some loss of DNA integrity, but growth was enhanced (Neuparth et al. 2005). Potential toxicants that might be responsible for those effects were identified by a weight of evidence method: Observed toxicity in sediment appeared to result from high copper levels. S2 seemed to be less toxic than S1, because it was further away from an industrial effluent at the sampling site and had a lower PAH concentration.

7.2.2 Sediment Assays Combined with Behavior

De Lange et al. (2006b) investigated the avoidance of PAH-contaminated sediments by two freshwater invertebrates, *G. pulex* and *A. aquaticus*. The aim of the study was to assess the effect of PAHs on habitat choice of the amphipod *G. pulex* and the isopod *A. aquaticus*. Therefore, clean field sediment was spiked with a mixture of four PAHs, fluoranthene, pyrene, chrysene, and benzo[*k*]fluoranthene, to a total concentration of 30 mg PAH/kg dry weight each. In laboratory experiments, both species were then offered a choice between PAH-spiked sediments and clean sediments. Results showed that both species avoided PAH-spiked sediment, whereas the origin of the population, either from a clean reference site or from a polluted site, did not affect habitat choice of either species.

7.3 *In Situ Tests*

7.3.1 Ecological Relevance of In Situ Data

Baird et al. (2007), in a review, examined how the choice of test species and study design employed in the use of in situ approaches in ecological risk assessment can maximize the ecological relevance of data. In situ effect measurement is a rapidly evolving field in ecotoxicology, with a variety of techniques employed to assess responses of ecosystems to toxic substances under field situations. Regarding "ecological relevance," Baird et al. (2007) and others (Crane et al. 2002) have stressed the value of in situ effect measures as a new and important line of evidence in ecological risk assessment. They provide a framework to define and assess ecological relevance and suggest that the following points should be taken into account for in situ study design:

1. *The test species*: If the in situ assay is used to generate information about effects on a resident biota within a specified area, a species relevant for the area of concern should be chosen (Pereira et al. 2000); moreover, the selected species should be abundant and play a key role in the local food web (Baird et al. 2007). However, when the aim of a study is to evaluate if the site in question is contaminated, a standard test organism may serve equally well. A standard organism may have a background data advantage. The disadvantage of a standard organism is that direct extrapolations to predict effects, within the local community, are not possible. Thus, ideally an "ecologically relevant test organism" should be used (Baird et al. 2007).
2. *The endpoints*: Baird et al. (2007) provide a working definition of an ecologically relevant effect: "It is a direct or an indirect response to exposure, resulting in changes in life-history characteristics that impair or diminish the growth potential of a population such that long-term maintenance of structural or functional qualities of that population, the community to which it belongs, or the ecosystem within which it exists is compromised or threatened".
3. *Relevance of effects*: Effects measured at one level of biological organization (i.e., feeding and metabolic rate) should be translatable into consequences for higher levels of organization (i.e., growth and reproduction), with ideally quantitative and mechanistic linkages to ensure robust extrapolations.
4. *Endpoints vs. protection goals*: Suborganism-level responses, i.e., biomarkers for stress, should be used to assess individual health and provide a useful early warning. Individual-level responses such as growth, behavior, reproduction, and survival should be used to assess individual performance. Population-level responses, for example, the growth of algal populations, should be used to assess population viability. Unfortunately, no animal-based population-level in situ bioassays exist to date. Finally, community-/ecosystem-level responses should be used to assess risks to biodiversity and ecosystem support services.
5. *Extrapolations*: To extrapolate from suborganism-level effects to higher-order consequences, the biomarkers utilized must be essential to normal function and be mechanistically linked to individual health status. To extrapolate from individual-level effects to population-level effects, information on stress-induced effects on vitality rates, measured in situ, can be combined with population models to predict population growth rate and abundance. Extrapolations from community-level effects with attributes of community structure have proven to be of value and can be assessed in situ (e.g., enclosures, recolonization, and transplantation). Such factors include species richness, organism abundance, biomass, and food web (trophic) composition. Similarly, community-level functional responses that are ecologically important can be assessed and include estimates of net primary productivity, carbon sequestration, and nutrient cycling.

6. *Using models*: To maximize the ecological relevance of data obtained from the use of in situ approaches (Baird et al. 2007).

7.3.2 In Situ Survival and Physiological Status

In situ exposures have also been used to assess effects of pH, humic substances, and animal interactions on survival and physiological status of *A. aquaticus* and *G. pulex* (Hargeby 1993). Caged *A. aquaticus* (Isopoda) and *G. pulex* (Amphipoda) were exposed for 25 days in 20 natural streams; these streams had a pH range of 4.3–7.5 and a color range of 8–280 mg Pt/L. In streams with a pH lower than 6.0, *G. pulex* responded with increased mortality and lower physiological status of surviving individuals. In *Asellus* the physiological status was correlated with pH and significantly affected by humus, whereas the mortality was not pH dependent (Hargeby 1993). Under optimal conditions of high pH and low humus concentrations, the interactions between the species appeared to be asymmetric, wherein the presence of *Gammarus* decreased the survival and physiological status of *Asellus*. The presence of *Asellus* did not increase the mortality or decrease the physiological status of *Gammarus*, because *Asellus* feed only on *Gammarus* that solely died from physiological stress. This mechanism suggests that food quality, and thus effects of diffuse competition, may be important in the context of withstanding acid stress. The results, though, give no support for the hypothesis that competition from *Asellus* is important for the disappearance of *Gammarus* as a consequence of stream acidification (Hargeby 1993).

7.3.3 In Situ Feeding Activity and Litter Breakdown

Feeding activity. Maltby et al. (2002) evaluated whether the *G. pulex* in situ feeding assay was useful for water quality biomonitoring. Uncontaminated reference sites were used to quantify background variability in feeding rates of aged *G. pulex* and to elucidate sources of variation. The ability of the assay to detect the impact of point-source discharges was assessed and the ecological relevance of the assay was determined by comparing assay responses to aspects of community structure and function. At the reference sites, feeding rate was strongly influenced by water temperature (76% of the variation), with a 30% feeding inhibition during the summer (>90% power). Downstream of point-source discharges, the inhibition of *G. pulex* feeding rates ranged between 27 and 99.6% (Maltby et al. 2002). These authors also found a strong positive correlation between in situ feeding rate, measured over 6 days, and leaf decomposition rate, measured over 28 days, as well as between in situ feeding and macroinvertebrate diversity and a biotic index. This underscored the importance of *G. pulex* as a detritivore in stream communities. Maltby et al. (2002) concluded that the *G. pulex* in situ feeding assay is a short-term sublethal biomonitor of water quality that is indicative of community- and ecosystem-level responses that occur over longer time periods. It is robust, responsive, and relevant.

Litter breakdown. Litter breakdown is a very suitable endpoint for assessing ecosystem conditions and the influence of multiple anthropogenic stresses imposed on them (Gessner and Chauvet 2002). Dangles et al. (2004) investigated the impact of stream acidification on litter breakdown and its consequences for assessing ecosystem function. Breakdown rates of the European Beech *Fagus sylvatica* varied more than 20-fold between the most acidified and the circumneutral sites, with stream water alkalinity and total Al concentration accounting for 88% of the variation in litter breakdown rates among streams. Interestingly, the abundance and biomass of the amphipod *G. fossarum*, an acid-sensitive and particularly efficient leaf shredder, showed a strong positive relationship with leaf breakdown rate. A variation of 85% in litter breakdown rates among streams could be accounted for by the combination of *G. fossarum* presence and microbial respiration (Dangles et al. 2004; Dangles and Guérold 2001).

The above results agree with results from Niyogi et al. (2001), wherein increased concentrations of zinc and increased deposition rates of metal oxides from mine drainage were closely related to a reduction in litter breakdown rates. The biomass of shredders was also found to decrease with decreased litter breakdown rates, whereby shredder biomass and microbial respiration together accounted for 76% of the variation in breakdown rates (Niyogi et al. 2001).

Microbial colonization and decomposition of leaves in a stream are also modified by coal ash effluent (Forbes and Magnuson 1981). Leaf surface area and disc weight were greater at the effluent-exposed site than at the reference site, after 96 days. Intercellular enzyme activity (measured by ATP content) of leaves from the reference stream quadrupled between 27 and 96 days, whereas ATP content of effluent-exposed leaves remained low. Macroinvertebrates colonized the leaf packs in the reference site but were not found on or in effluent-exposed packs, possibly as a consequence of reduced colonization and decomposition by fungi.

7.3.4 In Situ Tests as Part of a Whole Effluent Toxicity Study

A complete effluent toxicity (WET) study was conducted to investigate the toxicity and biological impact of a point-source discharge and to identify the in situ exposures to major toxicants of indigenous (*G. pulex*) and standard (*D. magna*). WETs are increasingly used to monitor compliance of consented discharges, but few studies have attempted to relate toxicity, measured using WET tests, to receiving water impacts. Maltby et al. (2000) adopted a four-stage procedure, in which the first stage standard WET tests were employed to determine the toxicity of the effluent, followed by the in situ deployment of *G. pulex* and *D. magna*. Then, biological survey techniques were used in the third stage to assess the impact of the discharge on the structure and functioning of the benthic macroinvertebrate community. Finally, in stage 4, toxicity identification evaluations (TIEs) were used to identify toxic components in the effluent. Maltby et al. (2000) found that receiving water toxicity and the ecological impact detected downstream of the discharge were consistent with the results of WET tests performed on the effluent. *D. magna* survival was reduced downstream of the discharge, as were survival and feeding rate of *G. pulex*. In addition, reductions in detritus processing and biotic indices, based on macroinvertebrate community structure, were found downstream of the discharge, most probably because of chlorine, which was determined by TIE studies to be the principal toxicant in the effluent (Maltby et al. 2000). With this approach, single species toxicity tests, community-level responses, and TIEs may be appropriately used to investigate effluent impacts.

7.3.5 In Situ Drift Behavior Resulting from Parasites

There is a question as to whether drift behavior of G. *pulex* is negatively affected by parasitism (McCahon et al. 1991). In a G. *pulex* population having approximately 20% *P. laevis*-infected adults, drift was monitored at margin and mid-river sites over a 24-h period. The authors observed a diurnal pattern of drift densities with a large increase at night, independent of parasite burden, and site location did not influence the proportion of parasitized and unparasitized G. *pulex* found in the drift or in the benthos. The drift of parasitized G. *pulex* was significantly greater than unparasitized animals and individuals harboring only one parasite were found in significantly higher proportions in the drift than were those with two or more parasites (McCahon et al. 1991). At both sites, significantly more unparasitized individuals were present in the benthos than in the drift, indicating that *P. laevis* infection alters drift behavior of G. *pulex* (McCahon et al. 1991).

8 Discussion

8.1 Evaluation of Existing Methods

The purpose of this review is to investigate the potential of gammarids as emerging test species for freshwater ecosystem effects (particularly in streams), by collecting available data, methods, and biomarkers on *Gammarus* spp. and then providing an overview of the currently used tests. We reviewed more than 200 publications that address the ecotoxicological effects on gammarids, biological background information, and related topics, and were surprised that so many aspects of gammarid ecotoxicology have already been investigated. The largest portion of the ecotoxicological investigations that have already been performed address acute toxicity testing (21 publications, but there are probably others not listed in Table 1). Lethal effect concentrations (LC_{50}s) were assessed for a diversity of environmental contaminants such as the metals cadmium, copper, zinc; the insecticides fenoxycarb and lindane; the pesticides esfenvalerate, terbutryn, and atrazine; and a herbicide (3,4-dichloroaniline). In all of the studies that were conducted either with juveniles or with adults for periods of 48–264 hour of exposure, juveniles appear to be more sensitive than adults; moreover, LC_{50} concentrations generally decrease with increasing exposure duration. Data on other invertebrates and fish species were also included in some of these. Generally, it appears that gammarids are among the most sensitive of exposed organisms. Compared with EC_{50} values found for dahpnids, gammarids were more sensitive toward esfenvalerate, lindane, atrazine, and copper. Thus, gammarids not only may be suitable as an additional invertebrate species for use in aquatic ecotoxicology testing, but also, in addition, are both a sensitive species and a stream-dwelling organism. Therefore, they cover a specific niche (streams of the Northern Hemisphere), which is not covered by planktonic daphnia found in lakes and ponds.

The second largest group of studies we reviewed addressed feeding activities of gammarids exposed to metals, antibiotics, insecticides, herbicides, and polluted water samples. Feeding activity proves to be a very sensitive endpoint for assessing general subacute toxicity, particularly for studies employing in situ exposures. One disadvantage may be the need for rather long exposure times when investigating low dose effects of contaminants. In addition, there may be difficulties in distinguishing between acute and chronic effects, and the nonspecific nature of some observed responses. Food choice and post-exposure feeding depression experiments may be good alternatives/extensions for feeding activity tests, because the former provides additional information on possible avoidance behaviors, whereas the latter gives an idea about reversible, sublethal effects after pulsed exposures.

Besides feeding, gammarid behavioral responses, such as locomotory and ventilatory activity, swimming endurance, pleopod beat frequency, and drift response, were also extensively studied. Behavioral responses are also rather nonspecific, when not linked with specific biomarkers, but proved to be very useful in in situ biomonitoring as a sensitive early warning endpoint. This is also the case for studies on population structure, which were employed in in situ comparisons of non-polluted and polluted sites.

In recent years, new studies have emerged on xenobiotics that address mode of action and identify new specific endpoints and biomarkers. For example, the effects of several known vertebrate xenoestrogens were investigated for possible endocrine activity in gammarids. The results indicate that some xenoestrogens possess endocrine potential in gammarids but do not necessarily target the same endpoints. Molting appears to be a central process that is at least partially controlled by the steroid hormone ecdysone, and it can be disrupted by ecdysone-like steroid hormones, as well as vitellogenin-inducing xenohormones. Other promising endpoints for xenoestrogens are sex ratio and gonad histology. Heat shock proteins or ecdysone have been investigated for their usefulness as biomarkers for crustacean endocrine disruption, but more data are needed to understand their roles. Moreover, other specific biomarkers such as stress response, metal exposure, oxidative stress, neurotoxicity, and detoxification have been used successfully for some time now.

In this review, we have summarized a diversity of test strategies with gammarids, ranging from acute to chronic in situ exposures and have covered a multitude of endpoints/effects, such as bioenergetic responses, metabolism, behavioral response, feeding activity, reproduction and population parameters, and in addition numerous specific biomarkers. In the future, the aim will be to combine and extend knowledge already gained in such areas to yield more integrative testing approaches with gammarids.

8.2 Perspectives on a Multimetric *Gammarus* spp. Test System

To date, few studies have incorporated sensitive, multi-stress test systems that use native invertebrate species for assessing freshwater ecosystem health in an integrated manner covering several biological organization levels and different levels of complexity and ecological relevance. Moreover, this gap is not yet filled by test systems proposed by international validation and standardization bodies.

We propose that *Gammarus* as an emerging test species for use in ecologically relevant and integrative aquatic ecotoxicity testing. *Gammarus* can be used in both in situ and ex situ testing, which allows one to link ecotoxicity test results to water quality bioassessment data for key species in European streams. Gammarid species hold a central position in the food web of streams, because they are structurally and functionally important keystone species; gammarids are also present in a variety of other ecosystems, ranging from freshwaters to marine waters. Such breadth of distribution provides opportunity for using them in different ecosystems as test organisms. Next to *D. magna/D. pulex* as representative invertebrate species for lakes, *Gammarus* is accepted in ecotoxicology for toxicity assessment in streams. *Gammarus* and *Daphnia* display similar sensitivities to toxic insult. Gammarids have frequently been used as test organisms in a suite of test methods for investigating different types of specific (MOA-based) and non-specific toxicity endpoints at different organizational levels, both in situ and in the lab.

Use of gammarids in integrative, multilevel elements of aquatic ecotoxicology studies requires that current methods be combined into more complex and sophisticated testing approaches; if properly done, such testing would be capable of assessing several different endpoints in one experiment. For example, the in vitro embryo culture method could be extended by incorporating biomarkers for endocrine disruption. To use *Gammarus* spp. as a standard test organism, it is necessary to improve the culturing methods for gammarids, which have recently been described by Bloor (2009). A steady supply of cultured, healthy embryos, juveniles, and adults is crucial for the development of new biomarkers. A key element in creating new biomarkers, e.g., for endocrine disruption, is to learn more of the sensitivity windows in the gammarid life cycle. With such insight, one can better distinguish between normal and pollutant-induced increases or decreases of selected biomarkers. The ability to maintain a gammarid culture is crucial if they are to become a widely accepted test species that can be used in other experimental setups (e.g., bioaccumulation, long-term, and multigenerational exposures to environmental micro-pollutants), thereby producing data with increased environmental relevance. Gammarids may also be useful for investigating possible relationships between molecular and biochemical biomarkers and endpoints on the physiological and organism level, for example, effects on behavior, histopathology, and morphology. Such operant biomarkers could become cost-effective tools to predict chronic long-term effects at the individual and population level.

9 Summary

The amphipod genus *Gammarus* is widespread and is structurally and functionally important in epigean freshwaters of the Northern Hemisphere. Its presence is crucial, because macroinvertebrate feeding is a major rate-limiting step in the processing of stream detritus. In addition, *Gammarus* interacts with multiple trophic levels by functioning as prey, predator, herbivore, detritivore, and shredder. Such a broad span of ecosystem participation underlines the importance of *Gammarus* spp. in freshwater ecosystems. The sensitivity of *Gammarus* to pollutants and other disturbances may render it a valuable indicator for ecosystem health.

This review summarizes the vast number of studies conducted with *Gammarus* spp. for evaluating aquatic ecotoxicology endpoints and examines the suitability of this native invertebrate species for the assessment of stream ecosystem health in the Northern Hemisphere. Numerous papers have been published on how pollutants affect gammarid behavior (i.e., mating, predator avoidance), reproduction, development, feeding activity, population structure, as well as the consequences of pollution on host–parasite, predator–prey, or native–invasive species interactions. Some biochemical and molecular biomarkers have already been established, such as the measurement of vitellogenin-like proteins, metallothioneins, alkali-labile phosphates (in proteins), and lipogenic enzyme activities for assessing endocrine disruption and detoxification mechanisms.

Despite the range and diversity of studies performed thus far on gammarids, we propose that future gammarid research should address the following aspects to enhance the integrative and multilevel approach outcomes in aquatic ecotoxicology testing:

1. Routine laboratory culturing and reproduction of gammarids seems to be difficult and is, to our knowledge, at the moment only done in one laboratory (Bloor 2009). As a pre-requisite for general use of gammarids in freshwater ecotoxicology, a solution is needed to this problem. Such a solution would allow testing of lab-cultured gammarids with different gender and age classes.
2. Up to the present, most chronic toxicity studies have been limited to short-term exposures. We propose that chronic in situ, ex situ, and pulsed long-term exposures be performed to investigate, for example, bioaccumulation, reproduction, and multigenerational effects of micro-pollutants. Such data are needed for incorporation into ecotoxicological risk assessment databases.
3. More information on the sensitivity windows is needed for existing and new gammarid biomarkers.
4. In future gammarid studies, it would be most useful to intensify the efforts to link biochemical, physiological, and molecular biomarkers to effects on behavior, histopathology, and morphology. Such a linkage would increase the ecotoxicological relevance of the data.
5. Efforts should be made to develop additional biomarkers. For example, in the field of endocrine disrupters, biomarkers are needed to enhance identification of

different EDC effects, such as effects on sexual development, reproduction, and molting. Such biomarkers would be useful in understanding the differences and similarities of endocrine disruption in invertebrates and vertebrates.

For future integrative ecotoxicology testing we suggest that established test procedures, for endpoints like feeding activity, behavior, development, and reproduction, be combined with new state-of-the-art, mode-of-action-based endpoints and biomarkers (i.e., for endocrine disruption and oxidative stress). Such a combination would produce an integrated, modular test system with *gammarids* for use in aquatic ecotoxicity testing. If available, this test system would fill a crucial gap in ecotoxicological assessments (e.g., as an ecotoxicological test system for addition to the water quality assessment within the EU-WaterFramworkDirective). It could become a sensitive, multilevel test system for use with a native invertebrate species for assessing in situ and ex situ freshwater ecosystem health in the Northern Hemisphere.

Acknowledgments We would like to thank Dave Whitacre for valuable and constructive comments and suggestions on the manuscript.

References

Abel PD (1980) Toxicity of γ-hexachlorocyclohexane (Lindane) to *Gammarus pulex*: Mortality in relation to concentration and duration of exposure. Freshwater Biol 10: 251–259.
Åbjörnsson L, Hansson A, Brönmark C (2004) Responses of prey from habitats with different predator regimes: Local adaptation and heritability. Ecology 85:1859–1866.
Adema DMM, Vink GJ (1981) A comparative study of the toxicity of 1,1,2-trichloroethane, dieldrin, pentachlorophenol and 3,4-dichloroaniline for marine and fresh water organisms. Chemosphere 10:533–554.
Allan J, Malmqvist B (1989) Diel activity of *Gammarus pulex* (Crustacea) in a South Swedish stream: Comparison of drift catches vs baited traps. Hydrobiologia 179:73–80.
Alonso A, Camargo JA (2004) Toxic effects of unionized ammonia on survival and feeding activity of the freshwater amphipod *Eulimnogammarus toletanus* (Gammaridae, Crustacea). Bull Environ Contam Toxicol 72:1052–1058.
Anderson NH, Cummins KW (1979) Influences of diet on the life histories of aquatic insects. J Fish Res Board Can 36:335–342.
Andersson K, Brönmark C, Herrmann J, Malmqvist B, Otto C, Sjörström P (1986) Presence of sculpins (*Cottus gobio*) reduces drift and activity of *Gammarus pulex* (Amphipoda). Hydrobiologia 133:209–215.
Arsuffi TL, Suberkropp K (1989) Selective feeding by shredders on leaf-colonising stream fungi: comparison of macroinvertebrate taxa. Oecologia 79:30–37.
Arthur JW, Leonard EN (1970) Effects of copper on *Gammarus pseudolimnaeus*, *Physa integra* and *Campeloma decisum* in soft water. J Fish Res Board Can 27:1277–1283.
Arts MJSJ, Schill RO, Knigge T, Eckwert H, Kammenga JE, Koehler HR (2004) Stress proteins (hsp70, hsp60) induced in isopods and nematodes by field exposure to metals in a gradient near Avonmouth, UK. Ecotoxicol 13:739–755.
Ashauer R, Boxall A, Brown C (2006) Predicting effects on aquatic organisms from fluctuating or pulsed exposure to pesticides. Environ Tox Chem 25:1899–1912.

Ashauer R, Boxall ABA, Brown CD (2007a) Modeling combined effects of pulsed exposure to carbaryl and chlorpyrifos on *Gammarus pulex*. Environ Sci Tech 41:5535–5541.

Ashauer R, Boxall ABA, Brown CD (2007b) Simulating toxicity of carbaryl to *Gammarus pulex* after sequential pulsed exposure. Environ Sci and Tech 41:5528–5534.

ASTM (1993) Standard guide for conducting sediment toxicity tests with freshwater invertebrates. In: Annual book of ASTM Standards, water and environmental technology, 11.04, ASTM, Philadelphia, E1383–E1393.

Atchison GJ, Henry MG, Sandheinrich MB (1987) Effects of metals on fish behavior: A review. Environ Biol Fish 18:11–25.

Baird DJ, Brown SS, Lagadic L, Liess M, Maltby L, Moreira-Santos M, Schulz R, Scott GI (2007) In situ-based effect measures: Determining the ecological relevance of measured responses. Integr Environ Assess Man 3:259–267.

Bakker TCM, Mazzi D, Zala S (1997) Parasite-induced changes in behavior and color make *Gammarus pulex* more prone to fish predation. Ecology 78:1098–1104.

Baldaia L, Porcheron P, Coimbra J, Cassier P (1984) Ecdysteroids in the shrimp *Palaemon serratus*: Relations with molt cycle. Gen Comp Endocriol 55:437–443.

Baldauf SA, Thünken T, Frommen JG, Bakker TCM, Heupel O, Kullmann H (2007) Infection with an acanthocephalan manipulates an amphipod's reaction to a fish predator's odours. Int J Parasitol 37:61–65.

Barlocher F, Kendrick K (1975) Assimilation efficiency of *Gammarus pseudolimnaeus* feeding on fungal mycelium or autumn shed leaves. Oikos 26:55–59.

Baumgärtner D, Jungbluth AD, Koch U, Von Elert E (2002) Effects of infochemicals on micro habitat choice by the freshwater amphipod *Gammarus roeseli*. Arch Hydrobiol 155:353–367.

Bayne BL, Moore MN, Widdows J, Livingstone DR, Salkeld PN (1979) Measurement of the responses of individuals to environmental stress and pollution: Studies with bivalve molluscs. Philos Trans R Soc Lond 286B: 563–581.

Beitinger TL (1990) Behavioral reactions for the assessment of stress in fishes. J Great Lakes Res 16:495–528.

Beketov M, Liess M (2008) Potential of 11 pesticides to initiate downstream drift of stream macroinvertebrates. Arch Environ Contam Toxicol 55:247–253.

Bermingham S, Dewey FM, Maltby L (1995) Development of a monoclonal antibody-based immunoassay for the detection and quantification of *Anguillospora longissima* colonizing leaf material. Appl Environ Microbiol 61:2606–2613.

Blockwell SJ, Taylor EJ, Jones I, Pascoe D (1998) The influence of fresh water pollutants and interaction with *Asellus aquaticus* (L.) on the feeding activity of *Gammarus pulex* (L.). Arch Environ Contam Toxicol 34:41–47.

Bloor M (2009) Aquatic pollution: Case study of landfill leachate toxicity and remediation. VDM Verlag, Germany pp. 1–176. ISBN 978-3-639-14699-8.

Bloor MC, Banks CJ (2006) An evaluation of mixed species *in-situ* and *ex-situ* feeding assays: The altered response of *Asellus aquaticus* and *Gammarus pulex*. Environ Int 32:22–27.

Bloor MC, Banks CJ, Krivtsov V (2005) Acute and sublethal toxicity tests to monitor the impact of leachate on an aquatic environment. Environ Int 31:269–273.

Bollache L, Cezilly F (2004) State-dependent pairing behaviour in male *Gammarus pulex* (L.) (Crustacea, Amphipoda): effects of time left to moult and prior pairing status. Behav Process 66:131–137.

Borowski B (1984) The use of the males' gnathopods during precopulation in some gammaridean amphipods. Crustaceana 47:245–250.

Bousfield EL (1973) Shallow-water gammaridean Amphipoda of New England. Corell University Press, Ithaca, New York.

Boxall ABA, Maltby L (1995) The characterization and toxicity of sediment contaminated with road runoff. Water Res 29:2043–2050.

Breneman DH, Pontasch KW (1994) Stream microcosm toxicity tests: Predicting the effects of fenvalerate on riffle insect communities. Environ Toxicol Chem 13:381–387.

Brown AF, Pascoe D (1989) Parasitism and host sensitivity to cadmium: An acanthocephalan infection of the freshwater amphipod *Gammarus pulex*. J Appl Ecol 26:473–487.

Brown VM (1968) The calculation of the acute toxicity of mixtures of poisons to rainbow trout. Water Res 2:723–733.

Brungs WA, Geckler JR, Gast M (1976) Acute and chronic toxicity of copper to the fathead minnow in a surface water of variable quality. Water Res 10:37–43.

Bundschuh M, Hahn T, Gessner MO, Schulz R (2009) Antibiotics as a chemical stressor affecting an aquatic decomposer-detritivore system. Environ Toxicol Chem 28:197–203.

Burton GA, Nelson MK, Ingersoll CG (1992) Freshwater benthic toxicity tests. In: Burton GA (ed) Sediment toxicity assessment. Lewis Publishers, Boca Raton, Florida, pp. 213–240.

Call DJ, Brooke LT, Kent RJ (1987) Bromacil and diuron herbicides: toxicity, uptake, and elimination in freshwater fish. Arch Environ Contam Toxicol 16:607–613.

Camus L, Olsen GH (2008) Embryo aberrations in sea ice amphipod *Gammarus wilkitzkii* exposed to water soluble fraction of oil. Mar Environ Res 66:221–222.

Cardoso AM, Barros CMF, Ferrer Correia AJ, Cardoso JM, Cortez A, Carvalho F., Baldaia L (1997) Identification of vertebrate type steroid hormones in the shrimp *Penaeus Japonicus* by tandem mass spectrometry and sequential product ion scanning. J Am Soc Mass Spectrom 8:365–370.

Carr RS, Linden O (1984) Bioenergetic responses of *Gammarus salinus* and *Mytilus edulis* to oil and oil dispersants in a model ecosystem. Mar Ecol Progr Ser 19:285–291.

Cezilly F, Gregoire A, Bertin A (2000) Conflict between co-occurring manipulative parasites? An experimental study of the joint influence of two acanthocephalan parasites on the behaviour of *Gammarus pulex*. Parasitology 120:625–630.

Chang CF, Jeng SR (1995) Isolation and characterization of the female-specific protein (vitellogenin) in mature female hemolymph of the prawn *Penaeus chinensis*. Comp Biochem Physiol B 112:257–263.

Cold A, Forbes VE (2004) Consequences of a short pulse of pesticide exposure for survival and reproduction of *Gammarus pulex*. Aquat Toxicol 67:287–299.

Correia AD, Costa FO, Neuparth T, Diniz ME, Costa MH (2001) Sub-lethal effects of copper-spiked sediments on the marine amphipod *Gammarus locusta*: Evidence of hormesis? Ecotoxicol Environ Res 4:32–38.

Correia AD, Lima G, Costa MH, Livingstone DR (2002) Studies on biomarkers of copper exposure and toxicity in the marine amphipod *Gammarus locusta* (Crustacea): I. Induction metallothionein and lipid peroxidation. Biomarkers 7:422–437.

Costa FO, Correia AD, Costa MH (1998) Acute marine sediment toxicity: A potential new test with the amphipod *Gammarus locusta*. Ecotoxicol Environ Saf 40:81–87.

Costa FO, Neuparth T, Correia AD, Helena Costa M (2005) Multi-level assessment of chronic toxicity of estuarine sediments with the amphipod *Gammarus locusta*: II. Organism and population-level endpoints. Mar Environ Res 60:93–110.

Crane M (1994) Population characteristics of *Gammarus pulex* (L.) from five English streams. Hydrobiologia 281:91–100.

Crane M, Delaney P, Watson S, Parker P, Walker C (1995) The effect of malathion 60 on *Gammarus pulex* (L.) below watercress beds. Environ Toxicol Chem 14:1181–1188.

Crane M, Maltby L (1991) The lethal and sublethal responses of *Gammarus pulex* to stress: Sensitivity and sources of variation in an in situ bioassay. Environ Toxicol Chem 10: 1331–1339.

Crane M, Sildanchandra W, Kheir R, Callaghan A (2002) Relationship between biomarker activity and developmental endpoints in *Chironomus riparius* Meigen exposed to an organophosphate insecticide. Ecotoxicol Environ Saf 53:361–369.

Crossland NO (1988) A method for evaluating effects of toxic chemicals on the productivity of freshwater ecosystems. Ecotoxicol Environ Saf 16:279–292.

Cummins KW, Klug MJ (1979) Feeding ecology of stream invertebrates. Ann Rev Ecol Syst 10:147–172.

Dangles O, Gessner MO, Guerold F, Chauvet E (2004) Impacts of stream acidification on litter breakdown: implications for assessing ecosystem functioning. J Appl Ecol 41:365–378.

Dangles O, Guérold F (2001) II. Leaf litter processing and invertebrates linking shredders and leaf litter processing: Insights from an acidic stream study. Int Rev Hydrobiol 86:395–406.

Dangles OJ, Guérold FA (2000) Feeding activity of *Gammarus fossarum* (Crustacea: Amphipoda) in acidic and low mineralized streams. Verh Int Ver Limnol 27:1–4.

De Coen WM, Janssen CR (2003) The missing biomarker link: Relationships between effects on the cellular energy allocation biomarker of toxicant-stressed *Daphnia magna* and corresponding population characteristics. Environ Toxicol Chem 22:1632–1641.

De Lange HJ, Lürling M, Van den Borne B, Peeters ETHM (2005) Attraction of the amphipod *Gammarus pulex* to water-borne cues of food. Hydrobiologia 544:19–25.

De Lange HJ, Noordoven W, Murk AJ, Lürling M, Peeters ETHM (2006a) Behavioural responses of *Gammarus pulex* (Crustacea, Amphipoda) to low concentrations of pharmaceuticals. Aquat Toxicol 78:209–216.

De Lange HJ, Sperber V, Peeters ETHM (2006b) Avoidance of polycyclic aromatic hydrocarbon-contaminated sediments by the freshwater invertebrates *Gammarus pulex* and *Asellus aquaticus*. Environ Toxicol Chem 25:452–457.

De Waal M, Poortman J, Voogt PA (1982) Steroid receptors in invertebrates. A specific 17β-estradiol binding protein in a sea star. Mar Biol Lett 3:317–323.

Dick JTA, Elwood RW (1996) Effects of natural variation in sex ratio and habitat structure on mate-guarding decisions in amphipods (Crustacea). Behav Process 133:985–996.

Dos Santos Carvalho C, Sellsue De Araujo HS, Fernandes MN (2004) Hepatic metallothionein in a teleost (*Prochilodus scrofa*) exposed to copper at pH 4.5 and pH 8.0. Comp Biochem Physiol B 137:225–234.

Fairchild JF, La Point TW, Zajicek JL, Nelson MK, Dwyer FJ, Lovely PA (1992) Population-, community- and ecosystem-level responses of aquatic mesocosms to pulsed doses of a pyrethroid insecticide. Environ Toxicol Chem 11:115–129.

Fairs NJ, Quinlan PT, Goad LJ (1990) Changes in ovarian unconjugated and conjugated steroid titers during vitellogenesis in *Penaeus monodon*. Aquaculture 89:83–99.

Felten V, Charmantier G, Mons R, Geffard A, Rousselle P, Coquery M, Garric J, Geffard O (2008) Physiological and behavioural responses of *Gammarus pulex* (Crustacea: Amphipoda) exposed to cadmium. Aquat Toxicol 86:413–425.

Fielding NJ, MacNeil C, Dick JTA, Elwood RW, Riddell GE, Dunn AM (2003) Effects of the acanthocephalan parasite *Echinorhynchus truttae* on the feeding ecology of *Gammarus pulex* (Crustacea: Amphipoda). J Zool 261:321–325.

Fitter R, Manuel R (1994) Collins photo guide to lakes, rivers, streams and ponds. Harper Collins, London.

Forbes A, Magnuson J (1981) Decomposition and microbial colonization of leaves in a stream modified by coal ash effluent. Hydrobiologia 76:263–267.

Ford AT, Fernandes TF, Robinson CD, Davies IM, Read PA (2006) Can industrial pollution cause intersexuality in the amphipod *Echinogammarus marinus*? Mar Pollut Bull 53: 100–106

Forget J, Livet S, Leboulenger F (2002) Partial purification and characterization of acetylcholinesterase (AChE) from the estuarine copepod *Eurytemora affinis* (Poppe). Comp Biochem Physiol C 132:85–92.

Forrow DM, Maltby L (2000) Toward a mechanistic understanding of contaminant-induces changes in detritus processing in streams: direct and indirect effects on detritivore feeding. Environ Toxicol Chem 19:2100–2106.

Fossi MC (1998) Biomarkers as diagnostic and prognostic tools for wildlife risk assessment: Integrating endocrine-disrupting chemicals. Toxicol Ind Health 14:291–309.

Friberg N, Andersen TH, Hansen HO, Iversen TM, Jacobsen D, Krojgaard L, Larsen SE (1994) The effect of brown trout (*Salmo trutta L.*) on stream invertebrate drift, with special reference to *Gammarus pulex* L. Hydrobiologia 294:105–110.

Fulton MH, Key PB (2001) Acetylcholinesterase inhibition in estuarine fish and invertebrates as an indicator of organophosphorus insecticide exposure and effects. Environ Toxicol Chem 20:37–45.

Gagné F, Blaise C (2002) Modulation of exoskeleton characteristics and elevation of vitellogenin by municipal contaminants in the brine shrimp *Artemia franciscana*. In: Aquatic toxicology workshop, Whistler, Canada.

Gagné F, Blaise C, Pellerin J (2005) Altered exoskeleton composition and vitellogenesis in the crustacean *Gammarus* sp. collected at polluted sites in the Saguenay Fjord, Quebec, Canada. Environ Res 98:89–99.

Gagou ME, Kapsetaki M, Turberg A, Kafetzopoulos D (2002) Stage-specific expression of the chitin synthase DmeChSA and DMeChSB genes during the onset of drosophila metamorphosis. Insect Biochem Mol Biol 32:141–146.

Garcia-de la Parra LM, Bautista-Covarrubias JC, Rivera-de la Rosa N, Betancourt-Lozano M, Guilhermino L (2006) Effects of methamidophos on acetylcholinesterase activity, behavior, and feeding rate of the white shrimp (*Litopenaeus vannamei*). Ecotoxicol Environ Saf 65:372–380.

Gerhardt A (1995) Monitoring behavioural responses to metals in *Gammarus pulex* (L.) (Crustacea) with impedance conversion. Environ Sci Pollut Res 2:15–23.

Gerhardt A (1996) Behavioural early warning responses to polluted water – Performance of *Gammarus pulex* L. (Crustacea) and *Hydropsyche angustipennis* (Curtis) (Insecta) to a complex industrial effluent. Environ Sci Pollut Res 3:63–70.

Gerhardt A (1999) Recent trends in online biomonitoring for water quality control. In: Gerhardt A (ed) Biomonitoring of polluted water, vol. 9. Trans Tech Publications Ltd, Zürich, pp. 95–118.

Gerhardt A (2007) Aquatic behavioral ecotoxicology – Prospects and limitations. Human Ecol Risk Assess 13:481–491.

Gerhardt A, Carlsson A, Ressemann C, Stich KP (1998) New online biomonitoring system for *Gammarus pulex* (L.) (Crustacea): In Situ test below a copper effluent in south Sweden. Environ Sci Technol 32:150–156.

Gerhardt A, Kienle C, Allan IJ, Greenwood R, Guigues N, Fouillac AM, Mills GA, Gonzalez C (2007) Biomonitoring with *Gammarus pulex* at the Meuse (NL), Aller (GER) and Rhine (F) rivers with the online Multispecies Freshwater Biomonitor$^{(R)}$. J Environ Monit 9:979–985.

Gerhardt A, Quindt K (2000) Waste water toxicity and bio-monitoring with *Gammarus pulex* (L.) und *Gammarus tigrinus* (Sexton) (Crustacea: Amphipoda). Wasser Boden 52:19–26.

Gerhardt A, Svensson E, Clostermann M, Fridlund B (1994) Monitoring of behavioral patterns of aquatic organisms with an impedance conversion technique. Environ Int 20:209–219.

Gessner MO, Chauvet E (2002) A case for using litter breakdown to assess functional stream integrity. Ecolog Appl 12:498–510.

Girling AE, Pascoe D, Janssen CR, PeitherA, Wenzel A, Schäfer H, Neumeier B, Mitchell GC, Taylor EJ, Maund SJ, Lay JP, Jüttner I, Crossland NO, Stephenson RR, Persoone G (2000) Development of methods for evaluating toxicity to freshwater ecosystems. Ecotoxicol Environ Saf 45:148–176.

Gowland BTG, Moffat CF, Stagg RM, Houlihan DF, Davies IM (2002) Cypermethrin induces glutathione *S*-transferase activity in the shore crab, *Carcinus maenas*. Mar Environ Res 54:169–177.

Graça MAS, Maltby L, Calow P (1993) Importance of fungi in the diet of *Gammarus pulex* and *Asellus aquaticus* I: feeding strategies. Oecologia 93:139–144.

Graça MAS, Maltby L, Calow P (1994) Comparative ecology of *Gammarus pulex* (L.) and *Asellus aquaticus* (L.) II: fungal preferences. Hydrobiologia 281:163–170.

Green DWJ, Williams KA, Pascoe D (1986) Studies on the acute toxicity of pollutants to freshwater macroinvertebrates. 4. Lindane ($\hat{1}^3$-hexachlorocyclohexane). Arch Hydrobiol 106:263–273.

Gross-Sorokin MY, Grist EPM, Cooke M, Crane M (2003) Uptake and Depuration of 4-Nonylphenol by the Benthic Invertebrate Gammarus pulex: How Important Is Feeding Rate? Environ Sci Technol 37:2236–2241.

Gross MY, Maycock DS, Thorndyke MC, Morritt D, Crane M (2001) Abnormalities in sexual development of the amphipod Gammarus pulex (L.) found below sewage treatment works. Environ Toxicol Chem 20:1792–1797.

Hans RK, Khan MA, Farooq M, Beg MU (1993) Glutathione-S-transferase activity in an earthworm (*Pheretima posthuma*) exposed to three insecticides. Soil Biol Biochem 25: 509–511.

Hargeby A (1993) *Asellus* and *Gammarus* spp. (Crustacea) in changing environments: Effects of acid stress and habitat permanence. In Department of Ecology/Limnology, Ph.D. University of Lund, Lund, Sweden.

Hartnoll RG, Smith M (1980) An experimental study of sex discrimination and pair formation in *Gammarus duebeni*. Crustaceana 38:253–264.

Hayes JD, Pulford DJ (1995) The glutathione S-transferase supergene family: Regulation of GST and the contribution of the isoenzymes to cancer chemoprotection and drug resistance. Crit Rev Biochem Mol Biol 30:445–600.

Hill IR, Matthiessen P, Heimbach F (1993) Guidance document on sediment toxicity tests and bioassays for freshwater and marine environments. SETAC-Europe Workshop on Sediment Toxicity Assessment (Renesse, NL, November 1993). SETAC-Europe, Brussels, Belgium. 105 pp.

Holomuzki JR, Hoyle JD (1990) Effect of predatory fish presence on habitat use and diel movement of the stream amphipod, *Gammarus minus*. Freshwater Biol 24:509–517.

Hynes HBN (1955) The reproductive cycle of some British freshwater Gammaridae. J Anim Ecol 24:352–387.

Jungmann D, Ladewig V, Ludwichowski K-U, Petsch P, Nagel R (2004) Intersexuality in *Gammarus fossarum* (Koch) A common inducible phenomenon? Arch Hydrobiol 159: 511–529.

Karaman G, Pinkster S (1977) Freshwater *Gammarus* species from Europe, North Africa and adjacent regions of Asia (Crustacea, Amphipoda). I. *Gammarus pulex* group and related species. Bijdr Dierk 47:1–97.

Kaushik NK, Hynes HBN (1971) The fate of dead leaves that fall into streams. Arch Hydrobiol 68:465–515.

Kinzler W, Kley A, Mayer G, Waloszek D, Maier G (2008) Mutual predation between and cannibalism within several freshwater gammarids: *Dikerogammarus villosus* versus one native and three invasives. Aquat Ecol 43:457–464.

Köhler HR, Zanger M, Eckwert H, Einfeldt I (2000) Selection favours low hsp70 levels in chronically metal-stressed soil arthropods. J Evol Biol 13:569–582.

Kosalwat P, Knight AW (1987) Acute toxicity of aqueous and substrate-bound copper to the midge, *Chironomus decorus*. Arch Environ Contam Toxicol 16:275–282.

Kregel KC (2002) Molecular biology of thermoregulation: Invited review: Heat shock proteins: modifying factors in physiological stress responses and acquired thermotolerance. J Appl Physiol 92:2177–2186.

Ladewig V, Jungmann D, Köhler HR, Schirling M, Triebskorn R, Nagel R (2006) Population structure and dynamics of *Gammarus fossarum* (Amphipoda) upstream and downstream from effluents of sewage treatment plants. Arch Environ Contam Toxicol 50: 370–383.

Lauridsen RB, Friberg N (2005) Stream macroinvertebrate drift response to pulsed exposure of the synthetic pyrethroid lambda-cyhalothrin. Environ Toxicol 20:513–521.

Lawrence AJ, Poulter C (1998) Development of a sub-lethal pollution bioassay using the estuarine amphipod *Gammarus duebeni*. Water Res 32:569–578.

Lawrence AJ, Poulter C (2001) Impact of copper, pentachlorophenol and benzo[a]pyrene on the swimming efficiency and embryogenesis of the amphipod *Chaetogammarus marinus*. Mar Ecol Progress Ser 223:213–223.

Lecerf A, Dobson M, Dang C, Chauvet E (2005) Riparian plant species loss alters trophic dynamics in detritus-based stream ecosystems. Oecologia 146:432–442.

Li GC, Werb Z (1982) Correlation between synthesis of heat shock proteins and development of thermotolerance in Chinese hamster fibroblasts. Proc Nat Acad Sci USA 79:3218–3222.

Lincoln RG (1979) British marine Amphipoda: Gammaridea. British Museum (Natural History), London, UK.

Livingstone DR (2001) Contaminant-stimulated reactive oxygen species production and oxidative damage in aquatic organisms. Mar Pollut Bull 42:656–666.

Macedo-Sousa JA, Gerhardt A, Brett CMA, Nogueira AJA, Soares AMVM (2008) Behavioural responses of indigenous benthic invertebrates (*Echinogammarus meridionalis*, *Hydropsyche pellucidula* and *Choroterpes picteti*) to a pulse of Acid Mine Drainage: A laboratory study. Environ Pollut 156:966–973.

Macek KJ, Buxton KS, Sauter S, Gnilka S, Dean JW (1976) Chronic toxicity of atrazine to selected aquatic invertebrates and fish. In: Ecol Res Ser. EPA-600/3-76-047. U.S. Environmental Protection Agency, Washington, DC.

Macneil C, Dick JTA, Elwood RW (1997) The trophic ecology of freshwater *Gammarus* spp. (Crustacea: Amphipoda): Problems and perspectives concerning the functional feeding group concept. Biol Rev 72:349–364.

Maitland PS (1966) Notes on the biology of *Gammarus pulex* in the River Endrick. Hydrobiologia 28:142–152.

Malbouisson JFC, Young TWK, Bark AW (1995) Use of feeding rate and re-pairing of pre-copulatory *Gammarus pulex* to assess toxicity of gamma-hexachlorocyclohexane (lindane). Chemosphere 30:1573–1583.

Maltby L (1992) Heterotrophic microbes. In: Calow P, Petts GI (eds) The rivers handbook. Blackwell Scientific Publications, Oxford, pp 165–194.

Maltby L, Clayton SA, Wood RM, McLoughlin N (2002) Evaluation of the *Gammarus pulex* in situ feeding assay as a biomonitor of water quality: robustness, responsiveness, and relevance. Environ Toxicol Chem 21:361–368.

Maltby L, Clayton SA, Yu H, McLoughlin N, Wood RM, Yin D (2000) Using single-species toxicity tests, community-level responses, and toxicity identification evaluations to investigate effluent impacts. Environ Toxicol Chem 19:151–157.

Maltby L, Crane M (1994) Responses of *Gammarus pulex* (Amphipoda, Crustacea) to metalliferous effluents: Identification of toxic components and the importance of interpopulation variation. Environ Pollut 84:45–52.

Maltby L, Hills L (2008) Spray drift of pesticides and stream macroinvertebrates: Experimental evidence of impacts and effectiveness of mitigation measures. Environ Pollut 156: 1112–1120.

Maltby L, Naylor C (1990) Preliminary observations on the ecological relevance of the *Gammarus* "scope for growth" assay: effect of zinc on reproduction. Funct Ecol 4:393–397.

Maltby L, Naylor C, Calow P (1990) Field deployment of a scope for growth assay involving *Gammarus pulex*, a freshwater benthic invertebrate. Ecotoxicol Environ Saf 19:292–300.

Marchant R, Hynes HBN (1981) Field estimates of feeding rate for *Gammarus pseudolimnaeus* (Crustacea: Amphipoda) in the Credit river, Ontario. Freshwater Biol 11:27–36.

Maund SJ, Taylor EJ, Pascoe D (1992) Population responses of the freshwater amphipod crustacean *Gammarus pulex* (L.) to copper. Freshwater Biol 28:29–36.

McCahon CP, Maund SJ, Poulton MJ (1991) The effect of the acanthocephalan parasite (*Pomphorhynchus laevis*) on the drift of its intermediate host (*Gammarus pulex*). Freshwater Biol 25:507–513.

McCahon CP, Pascoe D (1988a) Culture techniques for three freshwater macroinvertebrate species and their use in toxicity tests. Chemosphere 17:2471–2480.

McCahon CP, Pascoe D (1988b) Use of *Gammarus pulex* (L.) in safety evaluation tests: Culture and selection of a sensitive life stage. Ecotoxicol Environ Saf 15:245–252.

McIntosh AR, Peckarsky BL, Taylor BW (1999) Rapid size-specific changes in the drift of *Baetis bicaudatus* (Ephemeroptera) caused by alterations in fish odour concentration. Oecologia 118:256–264.

McLoughlin N, Yin D, Maltby L, Wood RM, Yu H (2000) Evaluation of sensitivity and specificity of two crustacean biochemical biomarkers. Environm Toxicol Chem 19:2085–2092.

McWilliam RA, Baird DJ (2002a) Application of postexposure feeding depression bioassays with *Daphnia magna* for assessment of toxic effluents in rivers. Environ Toxicol Chem 21: 1462–1468.

McWilliam RA, Baird D J (2002b) Postexposure feeding depression: A new toxicity endpoint for use in laboratory studies with *Daphnia magna*. Environ Toxicol Chem 21:1198–1205.

Mian LS, Mulla MS (1992) Effects of pyrethroid insecticides on nontarget invertebrates in aquatic ecosystems. J Agric Entomol 9:73–98.

Moore JW (1975) The role of algae in the diet of *Asellus aquaticus* L. and *Gammarus pulex* L. J Anim Ecol 44:719–730.

Mori K (1967) Histochemical study on the localization and physiological significance of glucose-6-phosphate dehydrogenase system in the oyster during the stage of sexual maturation and spawning. Tohoku J Agric Res 17:287–295.

Morritt D, Spicer JI (1996) The culture of eggs and embryos of amphipod crustaceans: Implications for brood pouch physiology. J Mar Biol Assoc UK 76:361–376.

Munawar M, Norwood WP, McCarthy LH, Mayfield CI (1989) In situ bioassessment of dredging and disposal activities in a contaminated ecosystem: Toronto Harbour. Hydrobiologia 188–189:601–618.

Musko IB, Meinel W, Krause R, Barlas M (1990) The impact of Cd and different pH on the amphipod *Gammarus fossarum* Koch (Crustacea: Amphipoda). Comp Biochem Physiol C 96.11–16.

Nadeau D, Corneau S, Plante I, Morrow G, Tanguay RM (2001) Evaluation for Hsp70 as a biomarker of effect of pollutants on the earthworm *Lumbricus terrestris*. Cell Stress Chap 6: 153–163.

Nakagawa Y, Nishimura K, Oikawa N, Kurihara N, Ueno T (1995) Activity of ecdysone analogs in enhancing N-acetylglucosamine incorporation into the cultured integument of *Chilo suppressalis*. Steroids 60:401–405.

Nation L (2002) Integument. In: Insect Physiology and Biochemistry. CRC Press, London, pp 89–115.

Naylor C, Maltby L, Calow P (1989) Scope for growth in *Gammarus pulex*, a freshwater benthic detritivore. Hydrobiologia 188–189:517–523.

Neuparth T, Correia AD, Costa FO, Lima G, Costa MH (2005) Multi-level assessment of chronic toxicity of estuarine sediments with the amphipod *Gammarus locusta*: I. Biochemical endpoints. Mar Environ Res 60:69–91.

Nilsson LM (1974) Energy budget of a laboratory population of *Gammarus pulex* (Amphipoda). Oikos 25:35–42.

Niyogi DK, Lewis WM, McKnight DM (2001) Litter breakdown in mountain streams affected by mine drainage: biotic mediation of abiotic controls. Ecolog Appl 11:506–516.

Oberdörster E, Rice CD, Irwin LK (2000) Purification of vitellin from grass shrimp *Palaemonetes pugio*, generation of monoclonal antibodies, and validation for the detection of lipovitellin in Crustacea. Comp Biochem Physiol C 127:199–207.

Olsen GH, Carroll J, Sva E, Camus L (2008) Cellular energy allocation in the Arctic sea ice amphipod *Gammarus wilkitzkii* exposed to the water soluble fractions of oil. Mar Environ Res 66:213–214.

Pantani C, Pannunzio G, De Cristofaro M, Novelli A A, Salvatori M (1997) Comparative acute toxicity of some pesticides, metals, and surfactants to *Gammarus italicus* Goedm. and *Echinogammarus tibaldii* Pink. and Stock (Crustacea: Amphipoda). Bull Environ Contam Toxicol 59:963–967.

Pascoe D, Kedwards TJ, Blockwell SJ, Taylor EJ (1995) *Gammarus pulex* (L.) feeding bioassay – Effects of parasitism. Bull Environ Contam Toxicol 55:629–632.

Pereira AMM, Soares AMVM, Goncalves F, Ribeiro R (2000) Water-column, sediment, and in situ chronic bioassays with cladocerans. Ecotoxicol Environ Saf 47:27–38.

Pratt WB, Toft DO (1997) Steroid receptor interactions with heat shock protein and immunophilin chaperones. Endocr Rev 18:306–360.

Ridley M (1983) The explanation of organic diversity. The comparative method and adaptations for mating. Oxford University Press, New York.

Rossi L (1985) Interactions between invertebrates and microfungi in freshwater ecosystems. Oikos 44:175–184.

Scheil V, Triebskorn R, Köhler HR (2008) Cellular and stress protein responses to the UV Filter 3-benzylidene camphor in the amphipod crustacean *Gammarus fossarum* (Koch 1835). Arch Environ Contam Toxicol 54:684–689.

Scherer E (1992) Behavioral responses as indicators of environmental alterations: approaches, results, developments. J Appl Ichthyol 8:122–131.

Schill RO, Görlitz H, Köhler HR (2003) Laboratory simulation of a mining accident: acute toxicity, hsc/hsp70 response, and recovery from stress in *Gammarus fossarum* (Crustacea, Amphipoda) exposed to a pulse of cadmium. BioMetals 16:391–401.

Schirling M, Jungmann D, Ladewig V, Ludwichowski KU, Nagel R, Köhler HR, Triebskorn R (2006) Bisphenol A in artificial indoor streams: II. Stress response and gonad histology in *Gammarus fossarum* (Amphipoda). Ecotoxicol 15:143–156.

Schirling M, Jungmann D, Ladewig V, Nagel R, Triebskorn R, Köhler HR (2005) Endocrine effects in *Gammarus fossarum* (Amphipoda): Influence of wastewater effluents, temporal variability, and spatial aspects on natural populations. Arch Environ Contam Toxicol 49: 53–61.

Schirling M, Triebskorn R, Köhler HR (2004) Variation in stress protein levels (hsp70 and hsp90) in relation to oocyte development in *Gammarus fossarum* (Koch 1835). Invert Reprod Deve 45:161–167.

Schlenk D, Colley WC, El-AlfyA, Kirby R, Griffin BR (2000) Effects of the oxidant potassium permanganate on the expression of gill metallothionein mRNA and its relationship to sublethal whole animal endpoints in channel catfish. Toxicol Sci 54:177–182.

Schmidt J (2003) Wirkung von Umweltchemikalien auf *Gammarus fossarum* – Populationsexperimente und individuenbasiertes Reproduktionsmodell. In: Fakultät Forst-, Geo- und Hydrowissenschaften. Institut für Hydrobiologie. PhD Thesis. Technische Universität Dresden, Dresden.

Segner H, Caroll K, Fenske M, Janssen CR, Maack G, Pascoe D, Schäfers C, Vandenbergh GF, Watts M, Wenzel A (2003) Identification of endocrine-disrupting effects in aquatic vertebrates and invertebrates: report from the European IDEA project. Ecotoxicol Environ Saf 54: 302–314.

Shaw G (1979) Prey selection by breeding dippers. Bird Study 26:66–67.

Sibley PK, Kaushik NK, Kreutzweiser DP (1991) Impact of a pulse application of permethrin on the macroinvertebrate community of a headwater stream. Environ Pollut 70:35–55.

Sih A (1992) Prey uncertainty and the balancing of antipredator and feeding needs. Am Nat 139:1052–1069.

Smyly WJP (1957) The life-history of the Bullhead or Miller's Thumb (*Cottus gobio* L.). Proc Zool Soc Lond 128:431–453.

Streit B, Kuhn K (1994) Effects of organophosphorous insecticides on autochthonous and introduced *Gammarus species*. Water Sci Technol 29:233–240.

Subramoniam T (2000) Crustacean ecdysteroids in reproduction and embryogenesis. Comp Biochem Physiol C 125:135–156.

Subramoniam T, Tirumalai R, Gunamalai V, Hoffmann KH (1999). Embryonic ecdysteroids in a mole crab, *Emerita asiatica* (Milne-Edwards). J Biosci 24:91–96.

Sunny F, Lakshmy PS, Oommen OV (2002) Rapid action of cortisol and testosterone on lipogenic enzymes in a fresh water fish *Oreochromis mossambicus*: Short-term *in vivo* and *in vitro* study. Comp Biochem Physiol B 131:297–304.

Sutcliffe DW (1993) Reproduction in *Gammarus* (Crustacea: Amphipoda): male strategies. Freshwat Forum 3:97–109.

Sutcliffe DW, Carrick TR, Willoughby LG (1981) Effects of diet, body size, age and temperature on growth rates in the amphipod *Gammarus pulex*. Freshwater Biol 11:183–214.
Sutcliffe DW, Hildrew AG (1989) Invertebrate communities in acid streams. Cambridge University Press, Cambridge.
Taylor EJ, Jones DPW, Maund SJ, Pascoe D (1993) A new method for measuring the feeding activity of *Gammarus pulex* (L.). Chemosphere 26:1375–1381.
Taylor EJ, Maund SJ, Pascoe D (1991) Toxicity of four common pollutants to the freshwater macroinvertebrates *Chironomus riparius* Meigen (Insecta: Diptera) and *Gammarus pulex* (L.) (Crustacea: Amphipoda). Arch Environ Contam Toxicol 21:371–376.
Thomas F, Renaud F, Derothe JM, Lambert A, Meeüs T, Cézilly F (1995) Assortative pairing in *Gammarus insensibilis* (Amphipoda) infected by a trematode parasite. Oecologia 104:259–264.
Thornton JW, Need E, Crews D (2003) Resurrecting the ancestral steroid receptor: Ancient origin of estrogen signaling. Science 301:1714–1717.
Triebskorn R, Adam S, Casper H, Honnen W, Pawert M, Schramm M, Schwaiger J, Köhler HR (2002) Biomarkers as diagnostic tools for evaluating effects of unknown past water quality conditions on stream organisms. Ecotoxicol 11:451–465.
Van Wijngaarden RPA, Cuppen, JGM, Arts, GHP, Crum SJH, Van den Hoorn MW, Van den Brink PJ, Brock TCM (2004) Aquatic risk assessment of a realistic exposure to pesticides used in bulb crops: A microcosm study. Environ Toxicol Chem 23:479–1498.
Varo I, Navarro JC, Amat F, Guilhermino L (2002) Characterisation of cholinesterases and evaluation of the inhibitory potential of chlorpyrifos and dichlorvos to *Artemia salina* and *Artemia parthenogenetica*. Chemosphere 48:563–569
Veerasingham M, Crane M (1992) Impact of farm waste on freshwater invertebrate abundance and the feeding rate of *Gammarus pulex* L. Chemosphere 25:869–874.
Viarengo A, Burlando B, Ceratto N, Panfoli I (2000) Antioxidant role of metallothioneins: a comparative overview. Cell Mol Biol 46:407–417.
Vigh DA, Dendinger, JE (1982) Temporal relationships of postmolt deposition of calcium, magnesium, chitin and protein in the cuticle of the Atlantic blue crab, *Callinectes sapidus* Rathbun. Comp Biochem Physiol A 72:365–369.
Watts MM, Pascoe D, Carroll K (2001) Survival and precopulatory behaviour of *Gammarus pulex* (L.) exposed to two xenoestrogens. Water Res 35:2347–2352.
Watts MM, Pascoe D, Carroll K (2002) Population responses of the freshwater amphipod *Gammarus pulex* (L.) to an environmental estrogen, 17α-ethinylestradiol. Environ Toxicol Chem 21:445–450.
Watts MM, Pascoe D, Carroll K (2003) Exposure to 17[alpha]-ethinylestradiol and bisphenol A–effects on larval moulting and mouthpart structure of *Chironomus riparius*. Ecotoxicol Environ Saf 54:207–215.
Webster JR, Benfield EF (1986) Vascular plant breakdown in freshwater ecosystems. Ann Rev Ecol System 17:567–594.
Welton JS (1979) Life-history and production of the amphipod *Gammarus pulex* in a Dorset chalk stream. Freshwater Biol 9:263–275.
Welton JS, Clarke RT (1980) Laboratory studies on the reproduction and growth of the amphipod, *Gammarus pulex* (L.). J Anim Ecol 49:581–592.
Welton JS, Mill CA, Rendle EL (1983) Food and habitat partitioning in two small benthic fishes, *Noemacheilus barbatulus* (L.) and *Cottus gobio* (L.). Arch Hydrobiol 97:434–454.
Werner I, Auel H, Friedrich C (2002) Carnivorous feeding and respiration of the Arctic under-ice amphipod *Gammarus wilkitzkii*. Polar Biol 25:523–530.
Whitehurst IT, Lindsey BI (1990) The impact of organic enrichment on the benthic macroinvertebrate communities of a lowland river. Water Res 25:625–630.
WHO (1986). Organophosphorus insecticides: A general introduction. Environ Health Crit 63. World Health Organization, Geneva.
Williams DD, Moore KA (1985) The role of semiochemicals in benthic community relationships of the lotic amphipod *Gammarus pseudolimnaeus*: a laboratory analysis. Oikos 44:280–286.

Willoughby LG, Earnshaw R (1982) Gut passage times in *Gammarus pulex* (Crustacea: Amphipoda) and aspects of summer feeding in a stony stream. Hydrobiol 97:105–117.

Willoughby LG, Sutcliffe DW (1976) Experiments on feeding and growth of the amphipod *Gammarus pulex* (L.) related to its distribution in the River Duddon. Freshwater Biol 6:577–586.

Winner RW, Farrell M (1976) Acute and chronic toxicity of copper to four species of Daphnia. J Fish Res Bd Can 33:1685–1691.

Wisenden BD, Cline A, Sparkes TC (1999) Survival benefit to antipredator behavior in the amphipod *Gammarus minus* (Crustacea: Amphipoda) in response to injury-released chemical cues from conspecifics and heterospecifics. Ethology 105:407–414.

Wisenden BD, Pohlman SG, Watkin EE (2001) Avoidance of conspecific injury-released chemical cues by free-ranging *Gammarus lacustris* (Crustacea: Amphipoda). J Chem Ecol 27:1249–1258.

Wogram J, Liess M (2001) Rank ordering of macroinvertebrate species sensitivity to toxic compounds by comparison with that of *Daphnia magna*. Bull Environ Contam Toxicol 67:360–367.

Wudkevich K, Wisenden BD, Chivers DP, Smith, RJF (1997) Reactions of *Gammarus lacustris* to chemical stimuli from natural predators and injured conspecifics. J Chem Ecol 23:1163–1173.

Xuereb B, Noury P, Felten V, Garric J, Geffard O (2007) Cholinesterase activity in *Gammarus pulex* (Crustacea Amphipoda): Characterization and effects of chlorpyrifos. Toxicol 236:178–189.

Zielinski D (1998) Life cycle and altitude range of *Gammarus leopoliensis* Jazdzewski & Konopacka, 1989 (Amphipoda) in south-eastern Poland. Crustaceana 71:129–143.

Zou E, Fingerman M (1997) Synthetic estrogenic agents do not interfere with sex differentiation but do inhibit molting of the cladoceran *Daphnia magna*. Bull Environ Contam Toxicol 58:596–602.

Zou E, Fingerman M (1999) Effects of estrogenic agents on chitobiose activity in the epidermis and hepatopancreas of the fiddler crab, Uca pugilator. Ecotoxicol Environ Saf 42:185–190.

The Svalbard Glaucous Gull as Bioindicator Species in the European Arctic: Insight from 35 Years of Contaminants Research

J. Verreault, G.W. Gabrielsen, and J.O. Bustnes

Contents

1	Introduction	78
2	Contaminant Levels and Patterns	79
	2.1 Legacy and Emerging Organochlorines	79
	2.2 Chiral Legacy Organochlorines	81
	2.3 Brominated Flame Retardants	81
	2.4 Hydroxyl- and Methylsulfonyl-containing Metabolites	82
	2.5 Per-fluorinated and Poly-fluorinated Alkyl Substances	83
	2.6 Trace Elements and Organometals	83
	2.7 Other Contaminants	84
3	Temporal Trends	84
	3.1 Legacy Organochlorines	84
	3.2 Brominated Flame Retardants	86
	3.3 Mercury	86
4	Factors Influencing Bioaccumulation	86
	4.1 Gender and Maternal Transfer	86
	4.2 Age	87
	4.3 Feeding Ecology and Trophic Levels	87
	4.4 Site-Specific Accumulation	88
5	Biomarkers of Biological and Ecological Responses and Effects	89
	5.1 Biotransformation Enzymes and Porphyrins	93
	5.2 Retinoids	93
	5.3 Hormones and Transport Proteins	94
	5.4 Basal Metabolism and Thermoregulation	97
	5.5 Immunity and Parasites	98
	5.6 Chromosomes and DNA	100

J. Verreault (✉)
Département des sciences biologiques, Université du Québec à Montréal, Succursale Centre-ville, Montréal, QC, H3C 3P8, Canada
e-mail: verreault.jonathan@uqam.ca

D.M. Whitacre (ed.), *Reviews of Environmental Contamination and Toxicology*,
Reviews of Environmental Contamination and Toxicology 205,
DOI 10.1007/978-1-4419-5623-1_2, © Springer Science+Business Media, LLC 2010

5.7 Egg Characteristics 101
5.8 Feather Growth 101
5.9 Reproductive Behaviors 102
5.10 Reproductive Endpoints and Survival 103
5.11 Threshold Effect Levels 104
5.12 Egg Intake Advisory 105
6 Recommendations on the Use of Avian Bioindicator Species in the Arctic 105
7 Summary 107
Appendix 109
References 112

1 Introduction

The Svalbard archipelago (Norway) of the European Arctic is an important sink for anthropogenic chemicals transported via atmospheric and oceanic currents from distant sites of production and use. Continuous environmental monitoring of organohalogen compounds, trace elements, and organometals, in a variety of biotic and abiotic samples from Svalbard, has disclosed that this remote arctic region is among the most polluted in the polar regions (de Wit et al. 2004; Gabrielsen 2007; Letcher et al. 2009). The first survey of contaminants in Svalbard wildlife samples was reported in 1972 (Bourne and Bogan 1972). In this study, alarmingly high levels of polychlorinated biphenyls (PCBs) and p,p'-dichlorodiphenyldichloroethylene (p,p'-DDE) (311 and 67 ppm, respectively) were discovered in the liver of a glaucous gull (*Larus hyperboreus*). When found, this gull was in convulsions in a breeding colony on Bear Island, the southernmost island in the Svalbard archipelago (74° 22'N, 19° 05'E) (Fig. 1). Despite the passage of more than 35 yr since the Bourne and Bogan (1972) report, abnormal behaviors of this top scavenger–predator are regularly being recorded in Svalbard during the hatching and chick-rearing period (Sagerup et al. 2009a; Strøm, H. personal communication 2009). Although levels of legacy organochlorines (OCs) in the Arctic have shown a general decline over the last three decades (de Wit et al. 2004), a growing body of evidence suggests that the health of Svalbard glaucous gulls has been adversely affected by their high body burden of an increasingly complex array of contaminants to which they are exposed. It has recently been reported that the population of glaucous gulls from Bear Island has decreased by nearly 65%, from 2000 breeding pairs in 1986 to 650 in 2006 (Strøm 2007). Significant declines of glaucous gull populations have also been observed in the Canadian Arctic (Gaston et al. 2009). It has been suggested that the physiological stress induced by contaminants, among other potential anthropogenic or natural stressors (e.g., predation pressure, climate change, habitat loss, pathogens, and food scarcity), may have contributed to this dramatic population decline in Svalbard. However, a causal link between contaminant exposure and adverse health impact in Svalbard glaucous gulls, or any other arctic wildlife species (Letcher et al. 2009), remains to be established.

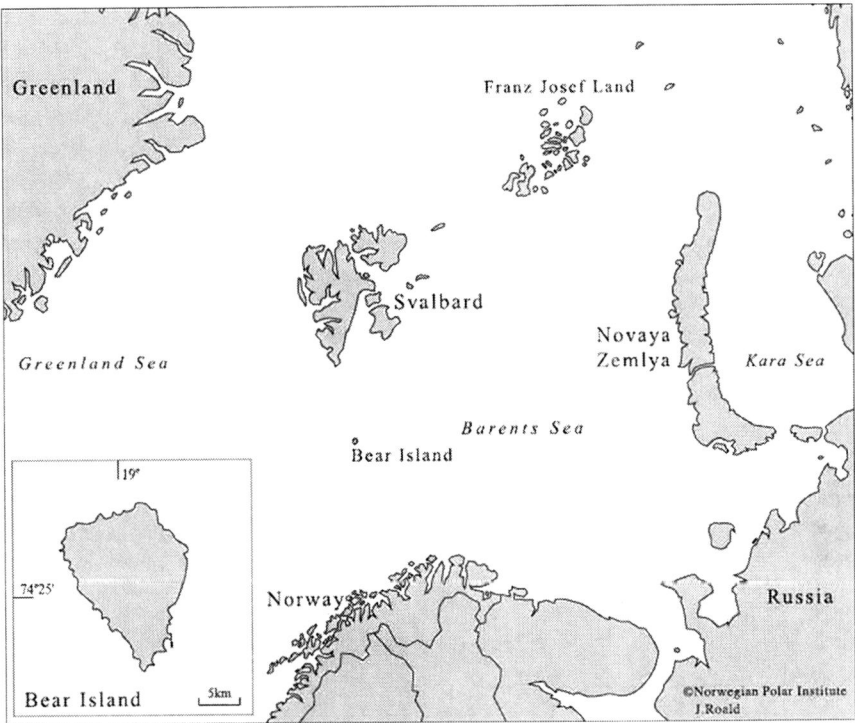

Fig. 1 Map of the European Arctic. Studies of glaucous gulls were conducted in Svalbard, primarily in the southern part of Bear Island, the southernmost island in this archipelago

In this chapter, it is our goal to comprehensively review and discuss the nearly 60 contaminant-related studies published over the last 35 yr that address contamination in the Svalbard glaucous gull. This review emphasizes contaminant levels and patterns, temporal trends in contamination, factors affecting bioaccumulation, as well as potentially useful biomarkers of biological and ecological responses and effects. We also provide recommendations on the use of glaucous gulls and other potentially useful apical avian species as bioindicator species in the European Arctic and elsewhere in the Arctic.

2 Contaminant Levels and Patterns

2.1 Legacy and Emerging Organochlorines

The legacy OCs, which encompass a large suite of chlorinated industrial chemicals, their byproducts, and pesticides, are undeniably the most studied classes of contaminants in tissues, blood, eggs, and intestinal content of Svalbard glaucous gulls – reports of contamination date from the early 1970s for PCBs and

p,p'-DDE (Bourne and Bogan 1972). In more recently collected samples from Svalbard, the blood plasma concentrations of the collective PCB congeners and dichlorodiphenyldichloroethane (DDT) compounds represent more than 72% of the total organohalogens found in this species (Fig. 2). Other legacy OCs that have routinely been detected in Svalbard glaucous gull studies, listed in order of decreasing abundance in plasma samples, include chlordane (CHL) compounds, chlorobenzene (CBz) congeners, dieldrin, hexachlorocyclohexane (HCH) isomers, and mirex (Fig. 2). Biosurveillance of legacy OCs in Svalbard glaucous gulls has also disclosed low levels of aryl hydrocarbon receptor (AhR)-inducible chemicals such as the coplanar non-*ortho* and mono-*ortho* PCB congeners, as well as the polychlorinated dibenzo-*p*-dioxin (PCDD) and polychlorinated dibenzofuran (PCDF) congeners (Daelemans et al. 1992; Henriksen et al. 2000; Pusch et al. 2005; Verreault et al. 2005a). More recently, lesser studied and/or emerging OCs have been detected in tissues, eggs, and plasma of Svalbard glaucous gulls, thus adding a novel dimension to the OC exposure profile in this species. These comprise the chlorobornane congeners (toxaphene) (Herzke et al. 2003; Verreault et al. 2005a) as well as the polychlorinated naphthalene (PCN) congeners, photo-mirex, pentachlorophenol (PCP), octachlorostyrene (OCS), and bis(4-chlorophenyl) sulfone

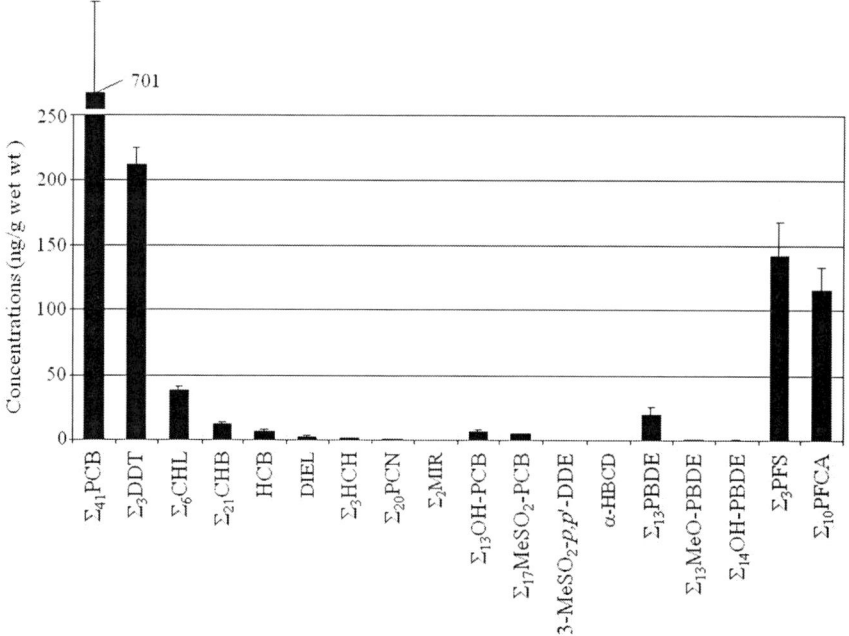

Fig. 2 Concentrations (ng/g wet wt) of 18 major organohalogen contaminant classes in blood plasma of breeding male glaucous gulls, sampled in 2002 and 2004, from Bear Island (Svalbard). Vertical bars indicate ±1 standard error of the mean. A definition of acronyms, sample size, and year of collection as well as a list of congeners and compounds included in these concentration sums can be found in Appendix. Data from Verreault et al. (2005a, 2005b, 2005c)

(BCPS) (Verreault et al. 2005a). With the exception of the chlorobornanes, these OCs were found to be of lesser bioaccumulative importance than the legacy OCs in Svalbard glaucous gulls. Hexachlorobutadiene (HCBD) (Verreault et al. 2005a) and short-, medium-, and long-chain chlorinated paraffin (SCCP, MCCP, and LCCP, respectively) congeners (Knudsen et al. 2007) have also been targeted for monitoring in this species (plasma, brain, and liver) but were consistently below the method detection limits.

2.2 Chiral Legacy Organochlorines

Several OC compounds display chirality and are present in the environment as pairs of enantiomers or atropisomers. Despite being released into the environment as racemic mixtures, non-racemic distributions of some chiral OCs have been detected in vertebrates and invertebrates, including various organisms of the arctic marine food web (Warner et al. 2005). The enantiomer fractions (EFs) of chiral CHL compounds, and atropisomeric PCB congeners detected in the blood plasma of breeding glaucous gulls from Svalbard, were reported to be identical in magnitude and character to those reported for Canadian Arctic glaucous gulls (Ross et al. 2008). These results strongly suggest that the biotransformation and/or dietary uptake processes that alter these stereochemical ratios are highly conserved among circumpolar glaucous gull populations. However, in the Ross et al. (2008) study, EFs determined in eggs (yolk) were seen to vary as a function of the colony from which they bred. After further examination, it was concluded that the female's specialization on certain food items (Section 4.3) may explain the variation in EFs between neighboring (∼2 km apart) Svalbard glaucous gull colonies.

2.3 Brominated Flame Retardants

Brominated flame retardants (BFRs) are widely used in consumer and industrial products to achieve fire safety standards; recent results demonstrate BFRs to be ubiquitous in the Arctic (de Wit et al. 2006). Herzke et al. (2003) were the first to report BFRs in Svalbard glaucous gulls. These authors monitored for a limited number of polybrominated diphenyl ether (PBDE) and polybrominated biphenyl (PBB) congeners. In their study, only two congeners (PBDE-47 and -99) were positively identified in liver and intestinal-content samples. Subsequent studies revealed the presence of a far more complex suite of tri- through deca-brominated PBDE congeners in Svalbard glaucous gull plasma, liver, and egg samples (Knudsen et al. 2005, 2006, 2007; Haukås et al. 2007; Verreault et al. 2005b, 2007a) (plasma; Fig. 2). In these surveys, concentrations of PBDE-47 largely dominated the overall PBDE profile, followed by nearly equal contributions from PBDE-99, -100, -153, and -154. Monitoring of PBDEs in Svalbard glaucous gull plasma and eggs also revealed the presence of PBDE-209 and its known or suspected octa- and nona-PBDE degradation products (e.g., PBDE-196, -197, -201, -202, -203, -205,

-206, -207, and -208) (Knudsen et al. 2006, 2007; Verreault et al. 2007a). However, these higher brominated congeners existed at lower levels than did the major PBDE congeners, often appearing at or near the instrumental limits of detection. These results suggest that PBDE-209 undergoes only limited debromination in the Svalbard glaucous gull or in its main prey items (e.g., fish, seal carcass, and seabird). At present, only two PBB congeners (PBB-101 and -153; the latter co-eluting with PBDE-154) have been detected in this species and these appeared at lower levels than did the PBDEs (Verreault et al. 2005b, 2007a). Moreover, several non-PBDE BFRs, which are alternatives or replacement products for the recently banned penta- and octa-PBDE mixtures, have been detected at low concentrations in eggs and plasma of Svalbard glaucous gulls (Verreault et al. 2007a). These BFRs include hexabromobenzene (HBB), 1, 2-bis (2, 4, 6-tribromophenoxy) ethane, pentabromoethylbenzene, pentabromotoluene, and α-hexabromocyclododecane (α-HBCD). In the Verreault et al. (2007a) investigation, α-HBCD concentrations were highest among the non-PBDE BFRs analyzed, followed by HBB (Verreault et al. 2007a). The β- and γ-HBCD isomers have not been detected in this species (liver and brain) (Knudsen et al. 2007). The results from these studies suggest that, in addition to a wide array of PBDE and PBB congeners, several current-use and as yet unregulated BFRs undergo long-range atmospheric transport and bioaccumulate at low levels in Svalbard glaucous gulls and are maternally transferred to the eggs of this species.

2.4 Hydroxyl- and Methylsulfonyl-containing Metabolites

The metabolism of PCBs and PBDEs via phase I (e.g., cytochrome P450 (CYP)) and phase II xenobiotic-metabolizing enzymes has been demonstrated to generate hydroxylated (OH) PCBs (OH-PCBs) and OH-PBDEs as well as methylsulfonate PCBs ($MeSO_2$-PCBs) in several vertebrate species (Hakk and Letcher 2003; Letcher et al. 2000). However, in addition to having a potential metabolic origin, certain OH-PBDEs (mainly *ortho*-OH-substituted congeners) and their methoxylated (MeO) analogues (MeO-PBDEs) have also been confirmed to occur naturally in the marine environment (Teuten et al. 2005). The OH-containing PCB and PBDE congeners were shown, in several animal models, to have higher bioactivity and toxicological potential (mainly endocrine disruption) than their respective parent compounds (Hakk and Letcher 2003; Letcher et al. 2000). A suite of OH-PCBs/-PBDEs, MeO-PBDEs, and $MeSO_2$-PCBs have been discovered in plasma of Svalbard glaucous gulls at concentrations substantially lower than their known or suspected PCB or PBDE precursors (Verreault et al. 2005a, 2005b) (Fig. 2). In eggs, OH-PCBs and $MeSO_2$-PCBs were detected at even lower concentrations, whereas OH-PBDEs were essentially nondetectable (Verreault et al. 2005a; Verreault, J. unpublished data). In these studies, the predominant OH-PCB and OH-PBDE congeners in plasma were 4-OH-PCB-187 and 6-OH-PBDE-47, respectively. The major $MeSO_2$-PCB congener in plasma was an unidentified hexa-chlorinated $MeSO_2$-PCB, whereas the dominant residues in eggs were the penta-chlorinated congeners 3′-$MeSO_2$-PCB-101 and 4′-$MeSO_2$-PCB-101. A $MeSO_2$-metabolite of

p,p'-DDE, 3-MeSO$_2$-p,p'-DDE, as well as a potential metabolite of OCS, 4-OH-heptachlorostyrene (4-OH-HpCS), were also detected in plasma and eggs of glaucous gulls (Verreault et al. 2005a). Despite their relatively low plasma levels, the presence of OH-PCBs and 4-OH-HpCS, and perhaps also certain OH- and MeO-PBDE congeners, supports the view that these are metabolically derived (CYP-mediated) in Svalbard glaucous gulls, whereas the MeSO$_2$-PCBs/-p,p'-DDE may also be dietary-sourced as a result of their bioaccumulation and biomagnification properties.

2.5 Per-fluorinated and Poly-fluorinated Alkyl Substances

Per- and poly-fluorinated alkyl substances (PFASs) have been used as surface-active agents in a multitude of manufactured and consumer products (e.g., fire-fighting foam and impregnation agent for carpets, papers, and textiles). The first survey of PFASs in Svalbard glaucous gulls, carried out by Verreault et al. (2005c), revealed relatively high concentrations of perfluorosulfonate (PFS) and perfluorocarboxylate (PFCA) compounds in plasma (Fig. 2), liver, brain, and egg samples. In this study, perfluorooctane sulfonate (PFOS) was the predominant PFS in all samples and was present at concentrations nearly comparable to the highly recalcitrant DDT compounds and PCB-153. Later surveys of PFASs confirmed these findings (Haukås et al. 2007; Knudsen et al. 2007), suggesting that PFOS should be regarded as an important bioaccumulative organohalogen in Svalbard glaucous gulls. In the Haukås et al. (2007) study, glaucous gulls were also found to accumulate the highest concentrations (in liver) relative to lower trophic-level species of the Barents Sea (Svalbard area) food web. These findings indicate that PFOS possess a high biomagnification potential in species from the arctic marine food web. Moreover, relatively high levels of PFCAs with 8–15 carbon (C) chains were reported in Svalbard glaucous gull samples, whereas 5C- and 6C-PFCAs were below the method limits of detection (Verreault et al. 2005c). In this study, the accumulation profile of PFCAs was characterized by high proportions of the long and odd-numbered C-chain length compounds, namely the perfluorodecanoic (11C) and perfluorotridecanoic (13C) acid, although their individual concentrations differed between plasma, tissues, and eggs. The following compounds could not be detected in any samples analyzed in this investigation: perfluorobutane sulfonate (PFBS), perfluorooctane sulfonamide (PFOSA), and four saturated (8:2 FTCA and 10:2 FTCA) and unsaturated (8:2 FTUCA and 10:2 FTUCA) fluorotelomer carboxylic acids.

2.6 Trace Elements and Organometals

The trace elements cadmium, zinc, lead, copper, selenium, and mercury were first analyzed in kidney and liver samples of Svalbard glaucous gulls collected in 1980 by Norheim and Kjos-Hanssen (1984) and Norheim (1987). These authors reported levels that were intermediate among other seabirds collected in the Svalbard area. The highest liver levels were found for zinc, followed by copper, cadmium, selenium, and mercury (lead was not detected) (Norheim 1987). A follow-up assessment of Svalbard glaucous gull muscle and liver samples collected a decade later, which also included arsenic but not lead, showed a somewhat consistent level profile: zinc >> copper > arsenic > cadmium > selenium > mercury (Savinov et al. 2003). Comparatively low levels of total mercury and methyl mercury were reported in more recently collected samples (liver, muscle, brain, and eggs) from this species (Jaeger et al. 2009; Knudsen et al. 2005, 2007; Sagerup et al. 2009a, 2009b). The work by Jæger et al. (2009) revealed a nonsignificant biomagnification tendency for total mercury and methyl mercury in organisms from the Svalbard marine food web, with the highest levels determined in glaucous gulls (muscle samples). Organometals, including organotins (mono-, di-, and tri-butyltins and -phenyltins), were also analyzed for in the liver of Svalbard glaucous gulls (Berge et al. 2004). In this study, only traces of di- and mono-butyltins were found in a few individuals, whereas the phenyltins were consistently below the method detection limits. In view of these results, trace elements and organometals appear to be of lower environmental concern in Svalbard glaucous gulls than are the major organohalogens and their metabolic products.

2.7 Other Contaminants

A limited number of non-halogenated and non-metallic compounds have been monitored for in Svalbard glaucous gull samples. Cyclododeca-1, 5, 9-triene, which is one potential degradation product of HBCD via reductive dehalogenation, was analyzed for, but not detected in eggs of this species (Knudsen et al. 2005). Interestingly, siloxane-D5, a cyclic siloxane having a wide range of commercial applications, was measured in liver of Svalbard glaucous gulls at concentrations nearly as high as some legacy OCs and PBDEs (Knudsen et al. 2007). This appears to be the first report of siloxane compounds in any arctic biota samples. It was concluded from this study that siloxane should be regarded as an emerging contaminant of potential health concern in Svalbard glaucous gulls. In this same screening study, low liver levels of anthracene (a tricyclic aromatic hydrocarbon) were also reported, whereas octylphenol was essentially nondetectable. Nonylphenol was detected, but only in the liver of four of the ten adult birds.

3 Temporal Trends

3.1 Legacy Organochlorines

Surveys of legacy OCs in blood (or plasma) of breeding glaucous gulls from Svalbard (Bear Island) have been performed on an annual or biannual basis between 1997 and 2006. Bustnes et al. (submitted) investigated the temporal changes of PCBs (sum of PCB-99, -118, -138, -153, -170, and -180), oxychlordane, and HCB

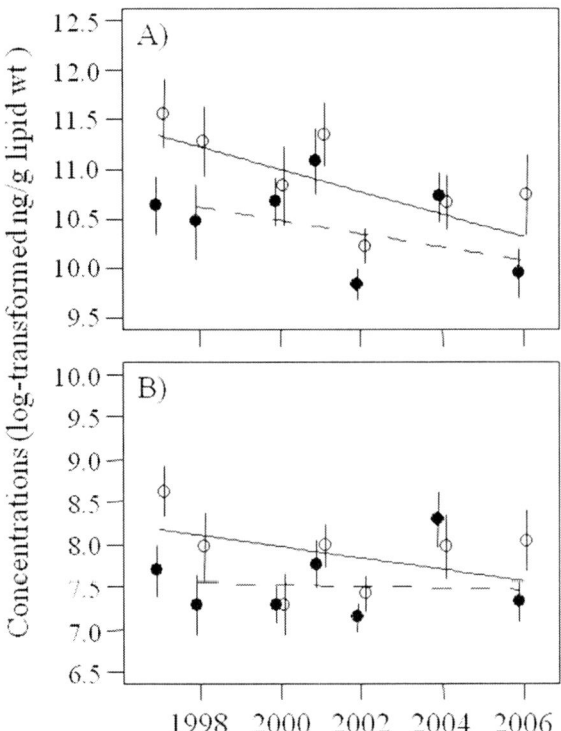

Fig. 3 Temporal trend of ΣPCB (sum of PCB-99, -118, -138, -153, -170, and -180) (**a**) and oxychlordane (**b**) concentrations (log-transformed ng/g lipid wt) in blood samples collected from male (*solid line, open circles*) and female (*stippled line, black circles*) glaucous gulls breeding on Bear Island (Svalbard) during the period 1997–2006. Data from Bustnes et al. (submitted)

concentrations in blood and plasma samples of 241 individuals collected during this 9-yr period (PCBs and oxychlordane; Fig. 3). Simple regression analyses showed that the concentrations of PCBs, oxychlordane, and HCB declined significantly (up to 60% for PCBs), particularly in males. Also examined in this study were the associations between legacy OC trends and selected biological parameters (sex and body condition) and climate variables, i.e., the Arctic Oscillation [AO] – a measurement of low-pressure activity and precipitation in the Arctic. The AO in the summer (June–September), prior to the sample collection period, had a positive effect on the concentrations of all compounds, whereas the winter AO (January–March) had a negative effect. This suggests that effects of increasing low-pressure activity and precipitation on POP (persistent organic pollutants) transport to the Arctic are very complex and depend on different timescales. The relationships between AO and POP accumulation may also have been confounded by AO effects on migration patterns in the glaucous gull and in its prey species. This temporal trend assessment suggests that although several legacy OCs are declining in Svalbard glaucous gull samples, environmental factors such as atmospheric activity may modulate the influx and subsequent food chain transfer of these compounds in the arctic ecosystem.

3.2 Brominated Flame Retardants

Eggs of breeding glaucous gulls were collected in Svalbard in 1997 (pooled sample) and 2002 and analyzed for eight major PBDE congeners and α-HBCD (Knudsen et al. 2005). This two-point temporal comparison did not reveal any significant change in egg PBDE concentrations, whereas those of α-HBCD were found to have increased markedly. No PBDE congener pattern change was observed during this 5-yr period.

3.3 Mercury

Knudsen et al. (2005) reported that levels of total mercury in eggs of Svalbard glaucous gulls were unchanged between samples collected in 1997 (pooled sample) and 2002.

4 Factors Influencing Bioaccumulation

4.1 Gender and Maternal Transfer

Notably higher organohalogen levels were reported in males than in females, in the preponderance of studies in which Svalbard glaucous gulls were sampled for blood shortly after clutch completion and up until hatching. This sex-specific difference results from the female's ability to transfer a portion of its lipophilic contaminant burden via egg formation, although dietary preference (e.g., seabird vs. fish intake) may also occur in this species (Bustnes et al. 2000). The dynamic of maternal transfer was investigated in Svalbard glaucous gulls for selected classes of lipophilic organohalogens including PCBs, DDTs, CHLs, CBzs, PBDEs, PBBs, and MeSO$_2$-PCBs (Verreault et al. 2006a). Organohalogen concentrations were determined in whole clutches, consisting of three eggs of known laying order, as well as in plasma collected from the respective laying females. Results from this study indicated that, in general, maternal transfer to eggs favors low K_{ow} and/or less persistent organohalogens, whereas the particularly recalcitrant and/or higher halogenated compounds are more selectively retained in the mother. However, the concentrations and compound mix patterns of most of the organohalogens determined in eggs fluctuated irrespective of their laying order in the clutch. Hence, for purposes of biomonitoring, it was concluded that Svalbard glaucous gull eggs collected randomly in a clutch would be representative of this clutch in terms of organohalogen levels and compound mix patterns. A complementary study by Ross et al. (2008) showed that the EFs of chiral CHL compounds and atropisomeric PCB congeners in eggs (yolk) and plasma of their respective mothers were highly consistent in Svalbard glaucous gulls. These results indicated that maternal transfer does not alter the stereochemical ratio between enantiomers of these chiral OCs. Based on this observation, it was concluded that egg yolk may also be used for biomonitoring of the enantiomeric distribution of chiral OCs in the Svalbard glaucous gull.

4.2 Age

A study of glaucous gulls breeding in Svalbard included an evaluation of the effect of gull age on blood concentrations of legacy OCs (Bustnes et al. 2003a). In this study, blood was collected from individuals that had been ring-marked as chicks or from individuals of unknown age that had been sampled repeatedly during four nonconsecutive years. The ages of the birds were not associated with blood concentrations of any of the analyzed OCs (PCB-153, p,p'-DDE, oxychlordane, and HCB). Moreover, the number of years that had elapsed between the sampling seasons in individuals of unknown age was not related to a variation in blood OC concentrations. These results indicate that, for glaucous gulls breeding in Svalbard, pseudo-steady-state OC levels are reached before or shortly after the age of first breeding (~5 yr)

4.3 Feeding Ecology and Trophic Levels

The influence of dietary specialization on blood concentrations of legacy OCs (PCBs, HCB, HCHs, oxychlordane, and p,p'-DDE) was investigated in two breeding colonies (~2 km apart) of Svalbard glaucous gulls (Bustnes et al. 2000). This study reported marked intercolony differences in OC concentrations that were associated with the bird's feeding ecology. Specifically, blood levels of OCs were highest in breeding gulls in the colony that preferred to feed on guillemot (*Uria* spp.) eggs and chicks, rather than those that fed predominantly on fish and crustaceans. The results of this study confirmed the importance of having a thorough knowledge of trophic feeding levels as the basis to explain intraspecific variation in OC levels in breeding Svalbard glaucous gulls. This phenomenon was further investigated by Sagerup et al. (2002), who examined the associations between the concentrations of a similar suite of legacy OCs in liver of breeding Svalbard glaucous gulls and their trophic levels. The trophic levels were estimated based on nitrogen ($^{15}N/^{14}N$: $\delta^{15}N$) and carbon ($^{13}C/^{12}C$: $\delta^{13}C$) isotope ratios in muscle and liver tissues. These authors reported that liver concentrations of HCB, p,p'-DDE, and five PCB congeners were positively correlated with levels of muscle tissue $\delta^{15}N$. No association was found between OC concentrations and $\delta^{13}C$ levels. These results suggested that OC accumulation in liver of Svalbard glaucous gulls can be only partially explained (up to 18% of the data variation) by their foraging strategy during the breeding season. Therefore, it was concluded that a better understanding of chemical bioavailability and toxicokinetic factors is required to predict OC levels in this species.

The results reported by Hop et al. (2002) corroborated the foregoing conclusion. These authors performed a comprehensive investigation of OC concentrations that existed in a variety of marine organisms (including glaucous gulls) that occupied different trophic levels (determined using $\delta^{15}N$) in the Barents Sea food web off Svalbard.

4.4 Site-Specific Accumulation

The earlier surveys of contaminants in Svalbard glaucous gulls routinely reported liver concentrations of legacy OCs (e.g., Bourne and Bogan 1972; Daelemans et al. 1992; Gabrielsen et al. 1995; Norheim and Kjos-Hanssen 1984; Savinova et al. 1995). However, since the late 1990s a non-destructive dosimetric method, based on the determination of OCs in blood samples (Bustnes et al. 2001a; Henriksen et al. 1998a), has generally been used to detect contaminants in glaucous gulls. Henriksen et al. (1998a) performed a study in which captive Svalbard glaucous gulls were fed a polar cod-based diet; results were that lipid-normalized blood concentrations of OCs were roughly equal to and positively correlated with those determined in liver. However, in this assessment, weak positive correlations were obtained when blood OC concentrations were compared with those measured in brain and subcutaneous adipose tissue samples. In another study, Bustnes et al. (2001a) repeatedly sampled Svalbard glaucous gulls within and between two consecutive breeding seasons. These authors showed that the variation in blood concentrations (wet wt) of the most persistent OCs, during one breeding season, could be easily predicted by those measured in the previous season. In a follow-up study, Verreault et al. (2007b) investigated a more comprehensive suite of organohalogens including legacy OCs (CHLs and PCBs) and BFRs (PBDEs, PBBs, and α-HBCD) as well as their metabolic products (OH-PCBs, OH-PBDEs, and $MeSO_2$-PCBs) in blood, liver, and whole body homogenate samples of captive Svalbard glaucous gulls fed an Arctic cod-based diet. In this study, the authors examined, in greater depth, the influence of the physicochemical properties of organohalogens and whole body composition (i.e., proportions of water, protein, lipid, and mineral contents) of breeding glaucous gulls on the variation of organohalogen levels. It was found that the proportions of OH-PCBs and OH-PBDEs to the total organohalogen concentrations were highest in blood, whereas the proportions of the lipophilic CHLs and PCBs were generally highest in liver and whole body homogenates. Moreover, the proportions of OH-PCBs, and to some degree PBDEs, were positively correlated with the total protein content isolated from the whole body homogenates of these birds. The total whole body homogenate lipid content was positively associated with PCB concentrations. It was recommended that both protein association and lipid solubility should be considered when investigating the toxicokinetics and fate of structurally different OH-substituted or OH-unsubstituted organohalogens. The authors concluded that a better understanding of site-specific accumulation of the more bioactive organohalogens (mainly endocrine disruptive ones), such as the

OH-PCB metabolites, is essential to understand their toxicological actions in the Svalbard glaucous gull.

5 Biomarkers of Biological and Ecological Responses and Effects

The bulk of effect studies with Svalbard glaucous gulls have disclosed relationships between organohalogen concentrations (blood or plasma) and various biomarkers of biological and ecological responses or effects. These biomarkers cover most of the organizational levels of the biological systems, from the molecular to the population level, and have been investigated in birds (mainly adults) from Bear Island during the incubation and chick-rearing period. A comprehensive summary of these relationships is presented in Table 1.

Table 1 Summary of relationships reported between contaminant concentrations and various biological and ecological responses or effects, in breeding Svalbard glaucous gull males (M) and females (F) (or combined sexes: M + F) and their eggs or chicks

Response or effect parameters	Relationships with contaminant concentrations[a]	Contaminants pinpointed based on the strength of the correlations[b]	References
Cytochrome P450 (CYP) enzymes and porphyrins			
CYP1A enzyme content (liver)	↑ (M); ↔ (F)	PCBs, DDTs, HCB	Østby et al. (2005)
Ethoxyresorufin-O-deethylase (EROD) activity (liver microsomes)	↑ (M + F) ↓ (M + F)	PCB-153 PCB-28, -47, -66, -74, -105, -187	Henriksen et al. (1998b, 2000) Henriksen et al. (1998c)
Testosterone hydroxylase (TH) activity (liver microsomes)	↔ (M + F)	–	Henriksen et al. (2000)
Highly carboxylated porphyrin (HCP) levels (liver)	↑ (M + F)	PCB-118	Henriksen et al. (2000)
Retinoids			
Retinol levels (liver)	↔ (M + F)	–	Henriksen et al. (1998c, 2000)
Retinyl palmitate levels (liver)	↔ (M + F)	–	Henriksen et al. (2000)
Hormones and binding proteins			
Free and total thyroxine (T_4) levels (plasma)	↓ (M); ↔ (F) ↔ (M + F)	HCB, oxychlordane –	Verreault et al. (2004) Verreault et al. (2007c)

Table 1 (continued)

Response or effect parameters	Relationships with contaminant concentrations[a]	Contaminants pinpointed based on the strength of the correlations[b]	References
Free and total triiodothyronine (T_3) levels (plasma)	$\leftrightarrow$ (M); $\leftrightarrow$ (F) $\leftrightarrow$ (M + F)	– –	Verreault et al. (2004) Verreault et al. (2007c)
Free and total T_4/T_3 ratios (plasma)	$\downarrow$ (M); $\leftrightarrow$ (F)	PCBs, p,p'-DDE, HCB, oxychlordane	Verreault et al. (2004)
	$\downarrow$ (M + F)	DDTs, CHLs, PCBs	Verreault et al. (2007c)
Total T_4/ free T_4 ratios (plasma)	$\downarrow$ (M); $\leftrightarrow$ (F) $\leftrightarrow$ (M + F)	oxychlordane –	Verreault et al. (2004) Verreault et al. (2007c)
Total T_3/ free T_3 ratios (plasma)	$\leftrightarrow$ (M); $\leftrightarrow$ (F) $\leftrightarrow$ (M + F)	– –	Verreault et al. (2004) Verreault et al. (2007c)
Testosterone (T) levels (plasma)	$\leftrightarrow$ (M); $\leftrightarrow$ (F)	–	Verreault et al. (2006b)
T levels (egg yolk)	$\uparrow$	MeO-PBDEs, CBzs, CHLs, HCHs, OCS	Verboven et al. (2008)
17β-estradiol (E_2) levels (plasma)	N.D.	–	Verreault et al. (2006b)
E_2 levels (egg yolk)	$\downarrow$	α-HBCD, PBDEs, PCBs, mirex, DDTs	Verboven et al. (2008)
Progesterone (P_4) levels (plasma)	$\uparrow$ (M); $\leftrightarrow$ (F)	PCBs, DDTs, CHLs, PBDEs	Verreault et al. (2006b)
Prolactin (PRL), baseline levels (plasma)	$\downarrow$ (M); $\leftrightarrow$ (F)	Legacy OCs, BFRs, OH-PCBs	Verreault et al. (2008)
PRL, post-handling levels (plasma)	$\leftrightarrow$ (M); $\leftrightarrow$ (F)	–	Verreault et al. (2008)
PRL, rate of decrease baseline-handling levels (plasma)	$\downarrow$ (M); $\leftrightarrow$ (F)	Legacy OCs, BFRs, OH-PCBs	Verreault et al. (2008)
Corticosterone (CORT), baseline levels (plasma)	$\uparrow$ (M); $\uparrow$ (F)	Legacy OCs, BFRs, OH-PCBs	Verboven et al. (2009a)
CORT, post-handling levels (plasma)	$\downarrow$ (M); $\leftrightarrow$ (F)	Legacy OCs, BFRs, OH-PCBs	Verboven et al. (2009a)
Organohalogen binding to recombinant transthyretin (rTTR) (liver and brain)	$\uparrow$ affinity than T_3 and T_4 (F)	OH-PCBs, OH-PBDEs	Ucán-Marín et al. (2009)
Basal metabolic rate (BMR) (*O_2 consumption*)	$\downarrow$ (M + F)	PCBs, DDTs, CHLs	Verreault et al. (2007c)

Table 1 (continued)

Response or effect parameters	Relationships with contaminant concentrations[a]	Contaminants pinpointed based on the strength of the correlations[b]	References
Nest temperature			
Normal clutch (three eggs)	↓ (M); ↓ (F)	PCBs, DDTs	Verboven et al. (2009b)
Artificially enlarged clutch (two to four eggs)	↔ (M); ↔ (F)	–	Verboven et al. (2009b)
Immune parameters			
Number of white blood cells (lymphocytes or heterophils)	↑ (M); ↑ (F)	PCBs, HCB, *p,p'*-DDE, oxychlordane	Bustnes et al. (2004)
	↓ (M); ↔ (F)	PCBs, DDTs	Verboven et al. (2009b)
Antibody response to tetanus toxoid	↔ (M); ↔ (F)	–	Bustnes et al. (2004)
	↔ (M + F)	–	Sagerup et al. (2009c)
Antibody response to diphtheria toxoid	↔ (M); ↑ (F)	HCB, oxychlordane	Bustnes et al. (2004)
Antibody response to influenza virus	↓ (M + F)	PCBs	Sagerup et al. (2009c)
Antibody response to reovirus	↔ (M + F)	–	Sagerup et al. (2009c)
Lymphocyte response to phytohemagglutinin-P	↑ (M + F)	PCBs	Sagerup et al. (2009c)
Lymphocyte response to pokeweed mitogen	↔ (M + F)	–	Sagerup et al. (2009c)
Lymphocyte response to concanavalin A	↑ (M + F)	PCBs	Sagerup et al. (2009c)
Immunoglobulin-G and -M levels (blood)	↓ (M + F)	PCBs	Sagerup et al. (2009c)
Parasite infection			
Nematode parasite burden (intestines)	↑ (M + F)	oxychlordane, DDTs, mirex, PCBs	Sagerup et al. (2000)
Parasite burden, several species (intestines)	↔ (M + F)	–	Sagerup et al. (2009a, 2009b)
Cestode and acanthocephalan parasite burden (intestines)	↑ (M + F)	selenium, mercury	Sagerup et al. (2009b)
Genetic damage			
Total number of chromosome aberrations (liver)	↔ (M); ↔ (F)	–	Krøkje et al. (2006)
Total number of damaged metaphases (liver)	↔ (M); ↔ (F)	–	Krøkje et al. (2006)
Total number of scorable metaphases (liver)	↔ (M); ↔ (F)	–	Krøkje et al. (2006)

Table 1 (continued)

Response or effect parameters	Relationships with contaminant concentrations[a]	Contaminants pinpointed based on the strength of the correlations[b]	References
Frequency of double-strand DNA breaks (liver)	↔ (M); ↔ (F)	–	Krøkje et al. (2006)
DNA adduct levels (liver)	↑ (M); ↑ (F)	–	Østby et al. (2005)
Egg characteristics			
Whole egg mass	↓	CHLs, α-HBCD	Verboven et al. (2009c)
Egg volume	↔	–	Bustnes et al. (2003b)
Egg yolk water content	↑	CBzs, PBDEs, OCS	Verboven et al. (2009c)
Egg albumen water content	↔	–	Verboven et al. (2009c)
Feather growth (probability of asymmetric wing feathers)	↑ (M + F)	PCBs, oxychlordane, p,p'-DDE, HCB	Bustnes et al. (2002)
Reproductive behaviors			
Proportion of time absent from the nest site	↑ (M + F)	PCBs, oxychlordane	Bustnes et al. (2001b, 2005)
Number of absences from the nest site	↑ (M + F)	PCBs	Bustnes et al. (2001b)
Proportion of time spent at the nest site	↔ (M); ↔ (F)	–	Verboven et al. (2009b)
Reproductive endpoints and survival			
Number of non-viable eggs	↑ (F)	PCBs, oxychlordane, p,p'-DDE, HCB	Bustnes et al. (2003b)
Egg-laying date	↓ (M); ↓ (F)	PCBs, oxychlordane, p,p'-DDE, HCB	Bustnes et al. (2003b)
Body condition of chicks at hatching	↑ (F)	PCBs, oxychlordane, p,p'-DDE, HCB	Bustnes et al. (2003b)
Length of incubation period	↔ (M); ↔ (F)	–	Bustnes et al. (2003b)
Clutch size	↔ (M); ↔ (F)	–	Bustnes et al. (2003b)
Adult survival	↓ (M); ↓ (F)	PCBs, oxychlordane, p,p'-DDE, HCB	Bustnes et al. (2003b, 2005)
Early chick growth	↔ (M); ↑ (F)	PCBs, oxychlordane, p,p'-DDE, HCB	Bustnes et al. (2005)
Early chick survival	↔ (M); ↔ (F)	–	Bustnes et al. (2003b)

[a] ↑: positive correlation; ↓: negative correlation; ↔: nonsignificant relationship
[b] A definition of chemical acronyms can be found in Appendix:
NA: data not available
ND: not detected

5.1 Biotransformation Enzymes and Porphyrins

The specificity, tissue content, and catalytic activity of xenobiotic-metabolizing enzymes are key factors that influence the fate and toxicokinetics of organohalogens in vertebrates. More specifically, the CYP isoenzymes within subfamilies 1, 2, and 3 play a central role in the phase I biotransformation of a broad variety of structurally diverse compounds (Hakk and Letcher 2003; Letcher et al. 2000). As a result, the content and/or catalytic activity of hepatic CYP isoenzymes has been widely used as a biomarker of contaminant exposure in wildlife species. Another biochemical marker that has been employed in free-ranging species is the level of highly carboxylated porphyrins (HCPs) in liver; these are indicative of chemically induced effects on the heme biosynthetic pathway (Fox et al. 2007). The content of CYP1A-like enzymes was measured in the liver of Svalbard glaucous gulls that had been exposed to a natural organohalogen-containing diet consisting of seabird eggs (exposed group) or a relatively clean diet composed of hen eggs (control group) (Østby et al. 2005). These researchers reported higher CYP1A enzyme levels in male glaucous gull liver from the exposed group than in controls, whereas no difference was found in females. Moreover, in males, CYP1A enzyme levels were positively correlated to the blood concentrations of most OCs (PCBs, DDTs, and HCB). Another biomarker studied in breeding Svalbard glaucous gulls is the catalytic activity of CYP1A-like enzymes, determined by using ethoxyresorufin-O-deethylase (EROD) activity in liver microsomes, or the liver concentrations of HCPs (uroporphyrin and hepta-, hexa-, and penta-carboxylic porphyrins) (Daelemans et al. 1992; Henriksen et al. 1998b, 1998c, 2000). In the studies by Henriksen and coworkers (1998b, 1998c, 2000), hepatic EROD activities and total HCP levels were reported to be low and showed inconsistent associations (i.e., positive, negative, or no correlation) with liver OC concentrations. Henriksen et al. (2000) also determined the activity of testosterone hydroxylase (CYP2B/3A-like enzyme activity) in breeding Svalbard glaucous gulls by quantifying the formation rates of the 6β-OH-testosterone metabolite in hepatic microsomes. In this investigation, the formation rates of 6β-OH-testosterone were not associated with the concentrations of any of the OCs analyzed. These authors concluded that the low catalytic activity of CYP-like isoenzymes and HCP levels, and their variable associations with the liver OC concentrations, suggest that these measures are poor biochemical markers of OC exposure in Svalbard glaucous gulls. It was further suggested that the Svalbard glaucous gull is fairly insensitive toward AhR-mediated effects of OCs as indicated by the low CYP1A-like enzyme activity.

5.2 Retinoids

Rolland (2000) suggested that OCs, particularly the dioxin-like compounds (e.g., PCDD, PCDF, and coplanar PCB congeners), may interfere with the homeostasis of vitamin A in wildlife species via modulation of enzymes involved in the metabolism of retinoids (retinol and retinyl palmitate). In studies of Svalbard glaucous gulls,

Henriksen et al. (1998c, 2000) showed that plasma retinoid levels were not associated with liver concentrations of OCs, although they were positively related to the hepatic microsomal EROD activity. However, in these same studies no association was found between EROD activity and liver OC concentrations. Based on these findings, the authors concluded that the measurement of liver retinoid levels is unsuitable as indicator of OC-mediated effects in the Svalbard glaucous gull.

5.3 Hormones and Transport Proteins

5.3.1 Thyroid Hormones and Transport Proteins

In several studies, altered circulating thyroid hormone (TH) levels and thyroid gland histology have been reported in free-ranging avian species exposed to organohalogens (McNabb 2005). The plasma status of total and free thyroxine (T_4) and triiodothyronine (T_3) was examined in Svalbard glaucous gulls nesting in two neighboring colonies: a low and a high OC-exposed colony (Verreault et al. 2004). These two breeding colonies were defined by previously documented differences in blood OC concentrations that resulted from the bird's preference for certain food items (Section 4.3). In this study (Verreault et al. 2004), breeding male glaucous gulls from the highest OC-exposed colony exhibited lower plasma levels of total and free T_4, and a tendency (not significant) for higher total and free T_3 levels, than existed in the less-exposed colony. Also, in males, negative correlations were reported between blood HCB and oxychlordane concentrations, and plasma total and free T_4 levels. Moreover, negative correlations were found in males between most OC (PCBs, HCB, p,p'-DDE, and oxychlordane) concentrations in blood and the free and total T_4- to T_3-level ratios (PCBs; Fig. 4). No significant result was found for any of these associations in females. The authors suggested that highly contaminated Svalbard male glaucous gulls, but apparently not the females, may be susceptible to changes in TH homeostasis. It was further suggested that chemically induced disruption in circulating TH levels may also affect basal metabolism and thermoregulation functions (Section 5.4). Nevertheless, in a follow-up study of breeding glaucous gulls sampled from proximal Svalbard colonies, somewhat different results were reported (Verreault et al. 2007c). In this assessment, in which males and females were combined because of low sample size, results indicated negative relationships (not significant) only between DDT, CHL, and PCB concentrations in plasma and ratios of free and total T_4 to T_3 levels. It was suggested that factors such as sample size and perhaps dietary composition (e.g., iodide content), nutritional status, age, activity level, and sensitivity of the thyroid system to organohalogens could have played a role in this apparent discrepancy. Regardless of these different, although minor, study outcomes, it was postulated in these studies that the TH status change observed in Svalbard glaucous gulls could be the consequence of competitive binding of certain organohalogens with the TH transport protein (e.g., transthyretin (TTR)) binding sites. Such binding affinity was investigated by Ucán-Marín et al. (2009), who isolated, cloned, and sequenced TTR

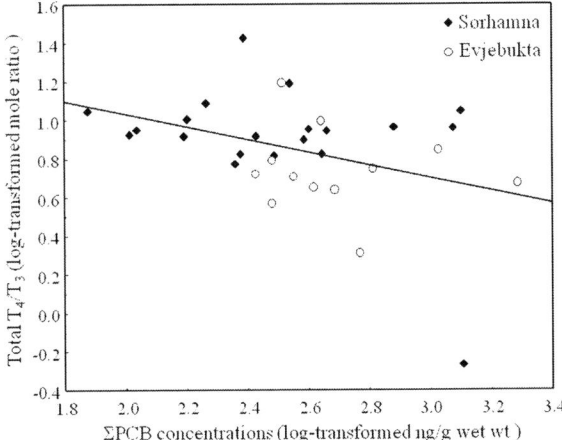

Fig. 4 Correlation (r = −0.40; p = 0.03) between total thyroxin to triiodothyronine level ratio ($T_4:T_3$; log-transformed mole ratio) and blood concentrations of ΣPCB (log-transformed ng/g wet wt), corrected for extractable plasma lipid percentages and day of capture in the incubation, for male glaucous gulls breeding in two colonies: a high organochlorine (OC)-exposed (Evjebukta) and a low OC-exposed colony (Sørhamna) on Bear Island (Svalbard). Figure from Verreault et al. (2004), reproduced with permission from Environ Health Perspect

cDNA from the brain and liver of a female glaucous gull from Svalbard. Competitive TTR binding was examined using recombinant TTR (rTTR) for differential concentrations of model PCB and PBDE congeners and their OH- and MeO-substituted analogues, as well as the natural TTR ligands, T_3 and T_4. PBDE-47, PCB-187, and their MeO-substituted analogues, and particularly the OH-compounds, all demonstrated lower affinity-constant (K_i) and dissociation-constant (K_d) values, indicating more potent competitive binding to both T_3 and T_4. It was suggested that OH-PCB congeners, and to a lesser extent OH-PBDE congeners, at concentrations that are currently found in Svalbard glaucous gulls (Verreault et al. 2005a, 2005b), have the potential to effectively displace T_4 and T_3 binding on TTR in this species.

5.3.2 Gonadal Steroid Hormones

Research has been performed on compounds that can mimic or block the actions of gonadal steroid hormones at the receptor level and to their interferences/interactions with the transport, biosynthesis, and metabolism of these hormones (Sanderson and van den Berg 2003). Circulating levels of testosterone (T), 17β-estradiol (E_2), and progesterone (P_4) were investigated in breeding Svalbard glaucous gulls in the context of their plasma levels of legacy OCs, BFRs, and the metabolites $MeSO_2$- and OH-PCBs (Verreault et al. 2006b). Results indicated positive correlations between plasma levels of P_4 and those of PCBs, DDTs, CHLs, and PBDEs in males, but not in females. The E_2 levels were below the radioimmunoassay detection limit in plasma of both males and females. No correlation was found between plasma organohalogen concentrations and those of T. The authors suggested that stimulation beyond the normal regulating processes of enzymes involved in P_4 synthesis and inhibition, or those that are responsible for P_4 metabolism, may have increased circulating P_4 levels in these male glaucous gulls. It was concluded that exposure to high organohalogen concentrations may interfere with steroidogenesis and impinge on circulating P_4 homeostasis in this species. In a complementary study, Verboven et al. (2008) reported positive correlations between the concentrations of MeO-PBDEs, CBzs, CHLs, HCHs, and OCS, and T levels in the yolk of unincubated, third-laid eggs of Svalbard glaucous gulls collected from one particular colony. No such relationship was found in the two other colonies that were investigated in this study. Changes in the relative concentrations of egg yolk E_2 levels also were observed as a function of the composition pattern of these organohalogens. It was concluded that contaminant-related modulation in the maternal transfer of gonadal steroid hormones to eggs may have occurred in glaucous gulls from this highly contaminated colony in Svalbard.

5.3.3 Prolactin

The anterior pituitary hormone prolactin is closely associated with reproduction in avian species and particularly with parental behavior (Buntin 1996). Contaminant-related changes in circulating levels of this hormone have rarely been investigated in wildlife. The variation of prolactin levels was assessed in breeding Svalbard glaucous gulls in the context of their plasma concentrations of legacy OCs, BFRs, and OH-metabolites (Verreault et al. 2008). Levels of prolactin were determined in plasma samples collected less than ~3 min following capture (baseline prolactin) of the gulls and after a 30-min standardized capture and restraint protocol (handling prolactin). The baseline prolactin levels and the rate of decrease in prolactin levels during this 30-min capture/restraint period tended to vary negatively with plasma concentrations of all the organohalogens analyzed in males, but not in females. Hence, the male glaucous gulls exhibited an attenuated stress-related prolactin response, which suggests that they maintained higher prolactin levels than the lower contaminated individuals. No correlation was found between plasma handling prolactin levels and organohalogen concentrations in either sex. It was concluded from this assessment that, in highly organohalogen-exposed Svalbard male glaucous gulls, the control of prolactin release may be affected by the direct or indirect modulating actions of contaminants and/or their metabolites. The altered baseline prolactin status in Svalbard male glaucous gulls was further conceived to be potentially associated with some of the adverse effects observed on their reproductive behaviors and development (Sections 5.8, 5.9 and 5.10).

5.3.4 Glucocorticoids

A short-term elevation of the circulating levels of corticosterone (the primary glucocorticoid in birds) in situations of acute environmental stress is considered to be an adaptive response that increases an individual's chances of survival and future reproductive success (Breuner et al. 2008). The levels of corticosterone were measured in plasma of breeding Svalbard glaucous gulls collected ~3 min after capture (baseline corticosterone) and following a 30-min standardized capture and restraint protocol (handling corticosterone) (Verboven et al. 2009a). High concentrations of legacy OCs, BFRs, and OH-PCBs/-PBDEs in the plasma of the captured glaucous gulls were associated with high baseline corticosterone levels in both sexes and a reduced stress response in males (that is, lower handling corticosterone levels). In view of these relationships, the authors concluded that the exposure to organohalogens potentially can increase the vulnerability of Svalbard glaucous gulls to other environmental stressors (e.g., food scarcity, predation, climate change, and pathogens). This apparent altered stress response was also thought to contribute, at least in part, to the low reproductive success and survival of Svalbard glaucous gulls (Sections 5.9 and 5.10).

5.4 Basal Metabolism and Thermoregulation

The basal metabolic rate (BMR) is commonly defined as the minimal rate of energy expenditure (or oxygen consumption) found in a thermoregulating, postabsorptive, adult endotherm at rest in its thermoneutral zone (Ellis and Gabrielsen 2002). In some studies, variations in BMR in free-ranging avian species underwent changes as thyroid activity (circulating T_3 levels) varied (Chastel et al. 2003). However, very limited work has been performed in birds to examine the concomitant effects of environmental contaminant exposure on basal metabolism and thyroid activity. Based on results from a previous study (Verreault et al. 2004), which showed that circulating TH homeostasis may be perturbed in organohalogen-exposed breeding glaucous gulls (males) in Svalbard, a follow-up assessment was conducted to investigate associations between total and free T_4 and T_3 levels, organohalogen concentrations, and BMR (Verreault et al. 2007c). Negative correlations were reported between BMR and plasma concentrations of PCBs, DDTs, and particularly CHLs in males and females, which were analyzed as one single group because of low sample size (sex was included as covariate). However, the plasma levels of total and free T_4 and T_3 were not associated with BMR changes or plasma concentrations of any studied organohalogens, although some tendencies emerged. It was concluded from this study that a depressed BMR could particularly affect highly contaminated gulls under repeated, stringent environmental conditions (e.g., low temperature) or stressful events (e.g., nest predation attempt). Moreover, altered BMR functions could result in insufficient energy allocated to reproduction in Svalbard glaucous gulls. Feeding rate (fat storage) and heat transfer to eggs and chicks (thermoregulation) may be among factors affected by energy allocation changes. Verboven et al. (2009b) probed the effects of organohalogen exposure on the thermoregulating capacity of Svalbard glaucous gulls from the same colonies studied by Verreault et al. (2007c). These authors explored the link between plasma organohalogen concentrations and thermal conditions inside the nest. Their study showed that, for normal clutches (three eggs), the nest temperature was negatively correlated with concentrations of many legacy OCs (PCBs and DDTs; Fig. 5), BFRs, and OH-PCB metabolites in plasma of the incubating male and female glaucous gulls. However, clutches in which the number of eggs was artificially augmented from two to four, thus increasing the energetic cost of incubation, displayed no association with plasma organohalogen concentrations (PCBs and DDTs; Fig. 5). These results suggest that fitness costs may accrue in the Svalbard glaucous gull through suboptimal thermal conditions during embryo development.

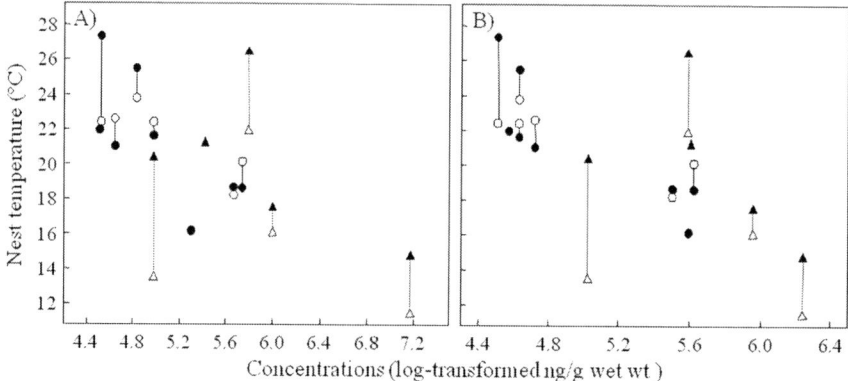

Fig. 5 Mean nest temperature for control two-egg clutches (*black symbols*) and experimentally enlarged four-egg clutches (*open symbols*), incubated by the same male (*triangle, stippled line*) or female (*circle, solid line*) glaucous gulls, from Bear Island (Svalbard) plotted against the concentrations (log-transformed ng/g wet wt) of ΣPCB (**a**) and ΣDDT (**b**) in plasma. Single black symbols represent three individuals that were observed on control clutches only. Figure from Verboven et al. (2009b), reproduced with permission from Anim Behav

5.5 Immunity and Parasites

5.5.1 White Blood Cells and Antibody Response

Various organohalogen contaminants are known to be immunotoxic to avian species, thus potentially increasing their susceptibility to infectious diseases (Grasman 2002). The relationships between blood (or plasma) concentrations of various organohalogens and variables relevant to the immune status and functions have been investigated in breeding Svalbard glaucous gulls and their chicks (Bustnes et al. 2004; Sagerup et al. 2009c; Verboven et al. 2009b). In the study by Bustnes et al. (2004), the number of white blood cells (lymphocytes and heterophils) was positively correlated with the blood concentrations of major legacy OCs (PCBs, HCB, p,p'-DDE, and oxychlordane) in both sexes. Interestingly, Verboven et al. (2009b) reported inverse correlations between plasma concentrations of PCBs and DDTs, and the number of white blood cells in Svalbard male glaucous gulls, whereas no correlation was found in females. Moreover, Bustnes et al. (2004) found evidence for decreased antibody response against the diphtheria toxoid, but not the tetanus toxoid (TET), in female glaucous gulls that had high blood levels of HCB and oxychlordane. However, no significant effect in males was observed, suggesting that their humoral immune response was not affected. Furthermore, in an experimental study design performed by Sagerup et al. (2009c), captive Svalbard glaucous gull chicks were fed for 8 wk on a naturally organohalogen-contaminated diet (seabird eggs) or a control diet (hen eggs). All chicks had been immunized with the herpesvirus (EHV), reovirus (REO), influenza virus (EIV), and the TET. Upon experiment completion, the chick group fed seabird eggs displayed liver concentrations of HCB, oxychlordane, p,p'-DDE, and PCBs that were 3- to 13-fold higher than those in the control group fed on hen eggs. The seabird-fed group had lower antibody response to the EIV and lower blood levels of immunoglobulin-G (IgG) and -M (IgM). Moreover, these same highly organohalogen-exposed chicks exhibited a higher peripheral blood lymphocyte response to phytohemagglutinin-P (PHA-P) and to spleen lymphocytes stimulated with concanavalin A (Con A) and PCBs. Based on these results, it was concluded that organohalogen (mainly legacy OCs) exposure in captive chicks and breeding glaucous gulls, in Svalbard, may affect the immune system of these birds, thus reducing immunocompetence and potentially their resistance to parasite infection (Section 5.5.2).

5.5.2 Parasite Infection

Factors that can suppress the immune system, which include exposure to environmental contaminants, may increase host susceptibility to parasite infection (Sures 2006). A study of breeding Svalbard glaucous gulls was conducted to investigate the relationships between intestinal macro-parasite (12 species of trematodes, cestodes, nematodes, and acanthocephalans) burden and liver concentrations of legacy OCs (Sagerup et al. 2000). In this study, individuals displaying high liver oxychlordane, DDT, mirex, and PCB levels also had a high intestinal nematode burden (four species combined) (Fig. 6). Although no immunological parameter was measured in this study, the data indicated that OC-contaminated Svalbard glaucous gulls may be less resistant to the establishment of intestinal parasites. By comparison, in subsequent studies performed by this same research group (Sagerup et al. 2009a, 2009b) on post-breeding Svalbard glaucous gulls, no correlation was found between parasite burden and liver levels of organohalogens (legacy OCs and BFRs). It was suggested that this discrepancy could result from differential reproductive status and body condition of the birds investigated. However, in the study by Sagerup et al. (2009b), in which trace elements were also analyzed in kidney, positive correlations were reported between intestinal cestode burden and selenium levels, and between intestinal acanthocephalan burden and mercury levels. In a related study, Bustnes et al. (2006) experimentally tested the interactions between legacy OC concentrations in blood, intestinal parasite burden, and various fitness components (nesting success and adult return rate between breeding seasons) in Svalbard glaucous gulls.

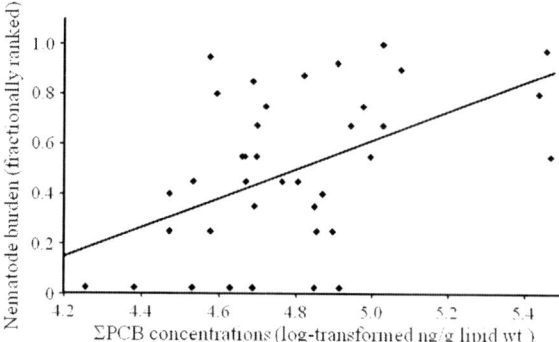

Fig. 6 Correlation ($r^2 = 0.26$; $p = 0.001$) between the number of intestinal nematodes (four species combined, fractionally ranked) and ΣPCB concentrations (log-transformed ng/g lipid wt) in breeding glaucous gulls (males and females combined) from Bear Island (Svalbard). Data from Sagerup et al. (2000)

In this investigation, the gulls were orally administered an antihelminthic drug or placebo solution during two consecutive breeding seasons. In males that received the placebo treatment, negative correlations were found between nesting success and blood OC concentrations, whereas in males treated with the anti-parasite drug there was no association between OC levels and these fitness components. These results suggest that enhanced parasite infestation in Svalbard glaucous gulls, for which the immune functions may have been altered via chronic OC exposure, may trigger important adverse reproductive effects and lower gull fitness.

5.6 Chromosomes and DNA

5.6.1 Chromosome Aberrations

Exposure to OCs, mainly PCBs, is known to induce genotoxic effects in some avian species (Dubois et al. 1995). The presence of chromosomal aberrations, quantified by cytogenetic analysis of blood cells, was investigated in captive Svalbard glaucous gull chicks fed hen eggs (control group) or naturally organohalogen-contaminated Svalbard seabird eggs (exposed group) (Krøkje et al. 2006). For both female and male chicks, the fraction of damaged metaphases was quantitatively higher (nonsignificant) in the exposed, relative to control groups. The number of aberrations per total number of damaged metaphase cells and the number of aberrations per total number of scorable metaphase cells were also tallied for the three categories of chromosomal damage (chromatid-interchange, -intrachange, and -break). These measures were higher, although not significantly, for all three categories in the exposed chick group compared to controls, with the exception of the number of breaks per damaged metaphase cell. The authors did not find any correlation between the frequency of chromosomal aberrations and the blood concentrations of OCs. The summary conclusion in this study was that dietary exposure to naturally organohalogen-contaminated seabird eggs did not induce noteworthy genotoxic effects in Svalbard glaucous gull chicks.

5.6.2 DNA Strand Break and Adduct Formation

The formation of DNA strand breaks and adducts was investigated in liver of captive Svalbard glaucous gull chicks fed hen eggs (control group) or naturally organohalogen-contaminated Svalbard seabird eggs (exposed group) (Krøkje et al. 2006; Østby et al. 2005). In the study by Krøkje and co-workers (2006), no difference was found between the exposed and the control chick groups in the frequency of double-strand DNA breaks. However, Østby et al. (2005) detected the formation of DNA adducts in livers of all birds, with the exception of one individual. The exposed chick group had significantly higher liver DNA adduct levels than did controls. No correlation was found between DNA adduct levels and blood concentrations of OCs. It was suggested that Svalbard glaucous gull chicks fed on environmentally contaminated seabird eggs may suffer from genetic damage by developing greater concentrations of DNA adducts in liver.

5.7 Egg Characteristics

It has previously been documented that exposure to some organohalogens (e.g., DDTs) can affect avian egg production, e.g., by reducing eggshell thickness (Moriarty et al. 1986). However, there is limited information on how contaminants affect other aspects of egg quality such as egg size and composition. A study designed to address such gaps was performed by Verboven et al. (2009c), who examined the relationships between organohalogen (legacy OCs, BFRs, and OH-metabolites) exposure in Svalbard glaucous gull females and various egg characteristics (egg mass, albumen mass, yolk mass, and yolk lipid, and water content). Glaucous gull females, with relatively high plasma concentrations of CHLs and α-HBCD, produced smaller eggs. It was also found that the compositional patterns (proportions of different compounds to their sums) of organohalogens in female glaucous gull plasma were associated with changes in water and lipid content in the egg yolk. Based on these findings, it was concluded that egg quality in Svalbard glaucous gulls may not only be affected by the direct transfer of contaminants from the mother to its eggs (Section 4.1) but also through changes in egg size and lipid and water content.

5.8 Feather Growth

Deviation from bilateral symmetry (non-directional or fluctuating asymmetry) is a morphological trait that has been used as a general indicator of environmental stress in wild birds (Clarke 1995). In a study of breeding Svalbard glaucous gulls, the association between wing feather asymmetry, i.e., the difference in length between the third primary feather of the left and the right wing, and blood concentrations of OCs was investigated (Bustnes et al. 2002). Positive correlations were reported between the probability of having asymmetric wing feathers and blood concentrations of two PCB congeners, oxychlordane, p,p'-DDE, and particularly HCB. In fact, at blood HCB levels above 30 ng/g wet wt, there was a 60% probability that the birds had asymmetric wing feathers. This study indicated that organohalogen exposure in glaucous gulls from Svalbard may cause developmental stress, which can then be reflected by higher wing feather asymmetry. It was suggested that feather growth, a process regulated mainly by the thyroid and gonadal steroid hormones, in glaucous gulls may be distorted by high circulating OC concentrations and may result from endocrine disruption (Sections 5.3.1 and 5.3.2).

5.9 Reproductive Behaviors

The behavioral effects of OC exposure that have been documented in wild birds include decreased parental attentiveness, impaired courtship behavior, and subtle neurological effects such as impaired avoidance behavior (Peakall 1996). Using two differentially exposed breeding colonies of glaucous gulls in Svalbard (Section 4.3), the relationship between measures of reproductive behavior (patterns of incubation and nest-site attentiveness) and blood concentrations of OCs was investigated (Bustnes et al. 2001b). After controlling statistically for the effect of colony and sex, the proportion of time absent from the nest site, when not incubating, and the number of absences from the nest site were both positively correlated to blood concentrations of PCBs. A follow-up study from this research group indicated that PCB and oxychlordane concentrations were positively correlated with the time away from the nest site, whereas p,p'-DDE and HCB concentrations were not related to this trait (Bustnes et al. 2005). By comparison, in another Svalbard glaucous gull study, in which reproductive behaviors were also monitored, no clear evidence was found to associate plasma concentrations of PCBs or DDTs with changes in the time a bird spent incubating (Verboven et al. 2009b). However, in this investigation, a nonsignificant tendency in the data set indicated that the proportion of time the males spent at the nest site was inversely related to plasma DDT concentrations. Overall, these studies were highly suggestive that Svalbard glaucous gulls retaining high blood OC concentrations exhibited an altered nest-site attendance and incubation behavior pattern. It was speculated that chemically related effects on circulating hormone levels (Section 5.3) or neurotoxicity may be involved in modulating reproductive behaviors in this species. Nest-site attendance and incubation behaviors are strongly influenced by hormonal fluctuations in birds. Therefore, concomitant effects of contaminants on circulating prolactin (Section 5.3.3), gonadal steroid (Section 5.3.2), and thyroid hormone homeostasis (Section 5.3.1) may occur and may explain, in part, the impaired reproductive behaviors of glaucous gulls. Abnormal reproductive behaviors were also thought to increase the birds' energetic costs during incubation

(Section 5.4), reduce their reproductive outputs (Section 5.10), and compromise nest defense against predators, including conspecifics and Arctic fox.

5.10 Reproductive Endpoints and Survival

Because of many uncontrolled factors, the assessment of fitness components (reproduction and survival) has been difficult to relate to contaminant exposure in wild birds (Hose and Guillette 1995). A study by Bustnes et al. (2003b) was designed to examine the relationships between various reproductive parameters (laying date, length of incubation period, clutch size, egg volume, hatching condition and survival of chicks, presence of non-viable eggs and adult survival) and levels of OCs in blood of Svalbard glaucous gulls. Results indicated that female glaucous gulls having the highest levels of the most persistent PCBs (higher chlorinated congeners), HCB, oxychlordane, and p,p'-DDE were more likely to lay non-viable eggs than lower exposed females. However, levels of the more volatile OCs (lower chlorinated PCB congeners and HCHs) were not related to egg viability. Moreover, negative correlations were reported between female blood concentrations for most of the persistent and more volatile OCs and the hatching condition of the chick from first-laid egg in the clutch and egg-laying dates. The hatching condition of the second chick from these females also was negatively related to blood concentrations of the higher chlorinated PCB congeners, HCB, oxychlordane, and p,p'-DDE. In a more comprehensive multiyear investigation, Bustnes et al. (2005) revisited the effect of blood OC levels on reproductive parameters in Svalbard glaucous gulls. The parameters included early chick growth and adult return rate from one breeding season to the following one. Early chick growth was negatively related to maternal blood levels of HCB, oxychlordane, p,p'-DDE, and PCBs in the low OC-exposed colony. In contrast, in females from the high OC-exposed colony, no relationship was found between blood OC levels and early chick growth. This suggests that natural or other anthropogenic stressors may be more important than thought in influencing OC-mediated effects. It was also reported that the probability of adult glaucous gulls returning to the breeding colonies from one year to the next one was lower in individuals having high blood levels of oxychlordane (Bustnes et al. 2003b, 2005) (Fig. 7). In fact, these authors believed that oxychlordane was the largest contributor to contaminant-induced toxicity in Svalbard glaucous gulls and played a primary role in the bird's lower return rate to the colonies and to their mortality, although the latter point remains to be verified. Moreover, in a recent study by Erikstad et al. (2009), the sex ratio of Svalbard glaucous gull chicks was examined in relation to their blood levels of legacy OCs. Glaucous gulls have male-biased size dimorphism and the sex allocation theory predicts that females under stress (e.g., high exposure to OCs) should skew the sex ratios of their chicks toward the less costly female offspring (Trivers and Willard 1973). Results were that healthy females with low OC concentrations produced more male chicks, whereas those in poor body condition produced more female chicks. However, unexpectedly, females with high levels of

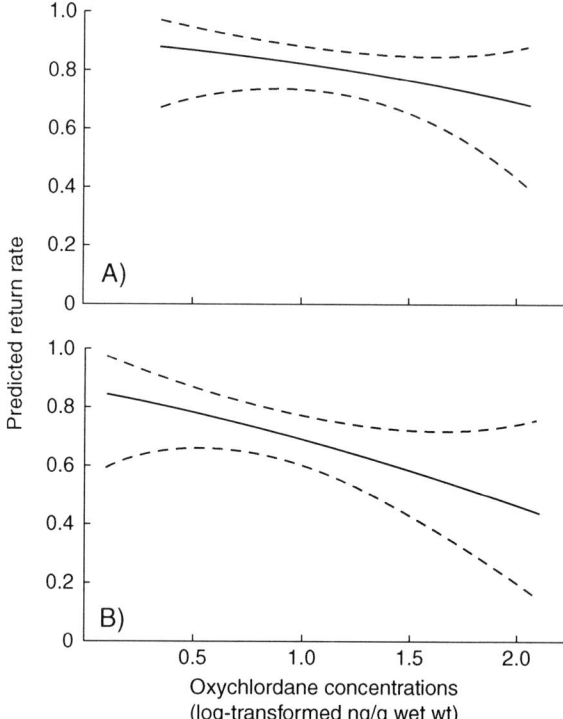

Fig. 7 Predicted adult return rate from 1997 to 1998, 2000 to 2001, and 2001 to 2002 of male (**a**) and female (**b**) glaucous gulls breeding on Bear Island (Svalbard) in relation to blood concentrations of oxychlordane (log-transformed ng/g wet wt). Figure from Bustnes et al. (2005), reproduced with permission from Environ Pollut

OCs produced more male chicks, and this trend was most evident among females in poor body condition. Furthermore, the body mass of male chicks at hatching was negatively related to female blood levels of OCs, although this association was significant only among chicks that hatched from the first egg in the clutch. It was suggested that the observed patterns could be the result of the capacity of certain OCs to disrupt the transport, biosynthesis, and/or metabolism of gonadal steroid hormones (Section 5.3.2) and hence influence the sex determining processes.

5.11 Threshold Effect Levels

Concentrations of the dioxin-like PCBs, measured in Svalbard glaucous gull liver by Daelemans et al. (1992), were compared to threshold levels associated with biological effects in various avian species (de Wit et al. 2004). In this comparative assessment, liver concentrations of mono-*ortho* PCBs, expressed as sum 2, 3, 7, 8-tetrachlorodibenzo-*p*-dioxin toxic equivalents (ΣTEQ; mean: 60.0 ng TEQ/g lipid wt) in a few glaucous gulls, were found to exceed the thresholds for no-observed-effects level (NOEL) for reproductive effects in black-crowned night herons (*Nycticorax nycticorax*) and Forster's terns (*Sterna forsteri*). Furthermore, some individuals exhibited ΣTEQ concentrations that were higher than the lowest-observed-adverse-effects level (LOAEL) of reproductive effects in common terns (*Sterna hirundo*), double-crested cormorants (*Phalacrocorax auritus*), and herring gulls (*Larus argentatus*). These reproductive effects comprised reproduction and hatching success, chick deformity, and egg mortality. Despite a substantially lower mean ΣTEQ level (25.4 ng TEQ/g lipid wt), reported by Henriksen et al. (2000) that was based on mono-*ortho* and non-*ortho* PCB concentrations in liver of Svalbard glaucous gulls, several of the birds also exceeded these NOEL and LOAEL. More recently, low ΣTEQ levels, calculated from the concentrations of mono-*ortho* and non-*ortho* PCBs as well as PCNs, were determined in Svalbard glaucous gulls. These ΣTEQ levels averaged 7.5 ng TEQ/g lipid wt in plasma of females and males combined and 12.5 ng TEQ/g lipid wt in whole eggs (Verreault et al. 2005a). Nevertheless, in a few birds of Verreault et al.'s (2005a) and Daelemans et al.'s (1992) studies, the lipid wt ΣTEQ in plasma and liver, respectively, was found to exceed the lowest-observed-effect level (LOEL; 25.0 ng TEQ/g lipid wt in liver) shown to induce hepatic CYP1A-like enzymes in common terns (de Wit et al. 2004).

5.12 Egg Intake Advisory

The human tolerable weekly intake (TWI) values calculated for Svalbard glaucous gull whole eggs, based on ΣTEQ of PCB, PCDD, and PCDF concentrations (mean: 0.11 ng TEQ/g wet wt), the accepted values established by the European Union Scientific Committee on Food (Pusch et al. 2005). Hence, considering the health risks associated with Svalbard glaucous gull egg intake, it was concluded from this study that children as well as young, pregnant, and nursing women should refrain from eating eggs from this arctic avian predator.

6 Recommendations on the Use of Avian Bioindicator Species in the Arctic

1. We recommend that research be concentrated to improve insights on the functioning of the biological systems and processes (e.g., immune, endocrine, reproductive, metabolic, etc.) before attempts are made to link biomarker responses to contaminant levels in species used or intended for use as bioindicators. One important confounding aspect in arctic wildlife population ecotoxicological research is the limited knowledge of and the general lack of control over several biological conditions and parameters. When the general biology of a species is poorly understood, the risk of erroneous study hypotheses and postulations increases, which could result in spurious health status assessments of chemically exposed wildlife populations. More knowledge is also needed concerning

species' behaviors (e.g., reproductive), feeding ecology, space-use strategy as well as life cycle events (phenology).

2. We recommend that the combined effects of multiple stressors be more deeply studied in arctic animals. Studies on arctic avian species exposed to environmental pollutants should consider the interacting effects of other anthropogenic or natural stressors. Emphasis should be placed on characterizing the most influential stressors in northern species and populations. Such stressors may include pathogen infection, food scarcity, predator pressure, habitat loss, and climate change. The latter point may be of great consequence. Recent study results have suggested that there is an additive or even synergistic effect from climate change on pollutant exposure-related effects. Both these factors retain high physiological stress momentum and may have far-reaching consequences on reproductive success and survival of arctic animals (Jenssen 2006). Moreover, Ims and Fuglei (2005) have predicted that changing climate will produce structural changes in species composition within arctic terrestrial and marine ecosystems. Changes in prey species composition (i.e., a shift toward alternative prey species), associated with a change in contaminant profile and level exposure, have recently been documented in Canadian Arctic polar bears (*Ursus maritimus*) (McKinney et al. 2009). Similarly, bird species that occupy high trophic levels in their respective terrestrial and marine food webs, and exhibit high dietary specialization, may be at higher risk for climate-associated changes in food web structure. The introduction of new more southern species or population increase of other less abundant species in the arctic regions as a result of the gradually changing Arctic climate may also increase the competitive interactions among species and populations for resources. Increased competition may have profound effects on the animal's space-use strategy (e.g., home range size modification). Examples of apex arctic avian species that could be at higher risk from concomitant contaminant exposure and climate change, and for which exceedingly little research has been done, are the ivory gull (*Pagophila eburnea*) and the great skua (*Stercorarius skua*). The ivory gull was recently (April 2006) listed as an endangered species by the Committee on the Status of Endangered Wildlife in Canada (COSEWIC; www.cosewic.gc.ca).

3. We recommend that work to identify and monitor for novel classes of contaminants in arctic wildlife species be emphasized. Moreover, high priority in such research should be given to top predator birds that represent ideal bioindicators species of environmental pollution in the arctic ecosystem. High trophic level birds in the Arctic are continuously exposed in their ecosystem to occasionally elevated concentrations of legacy and more recently introduced anthropogenic contaminants transported from distant sites of production and or use. Many of these emerging contaminants possess physicochemical properties that render them potential PBT (persistent, bioaccumulative, and toxic) candidates and endocrine-disrupting substances. Monitoring should include metabolic transformation products (e.g., OH-PCBs and OH-PBDEs) because these may exhibit higher toxicological potential (bioactivity) than their precursors/parents in vertebrates.

4. We recommend that biomarker responses and/or biological effects, observed in naturally contaminated arctic avian populations, be studied under controlled laboratory conditions and in semi-field scale studies. Moreover, special attention should be given to the study of other potential environmental and biological factors (e.g., pathogen infection, food scarcity, predator pressure, habitat loss, and climate change) and their implications on contaminant exposure during sensitive periods of the animal's life cycle (development, growth, and fasting). The use of laboratory-raised individuals of the same bird species or relevant surrogate (phylogenetically related) species is preferred (all ethical issues being addressed). It is further recommended that the performed experiments be based on exposure to environmentally representative pollutant mixtures reflecting the actual concentrations found in the main prey species under study. Well-designed studies will better establish cause–effect linkages and modes/mechanisms of action, and ultimately lead to the creation of predictive models of pollutant exposure and (threshold) effects in arctic avian wildlife.

7 Summary

Biomonitoring surveys conducted with glaucous gulls from Svalbard have demonstrated that this top predator–scavenger species accumulates a wide array of chemicals of environmental concern, including organohalogens, trace elements, organometals, and several non-halogenated and non-metallic compounds. Among these contaminants are those subjected to global bans or restrictions in North America and Europe (e.g., legacy OCs, penta- and octa-PBDE technical mixtures, and mercury). In addition, some currently produced chemicals were found in gulls that lack any global use regulation (e.g., deca-PBDE, HBCD, and other non-PBDE BFR additives, siloxanes, and selected PFASs). Svalbard glaucous gulls are also exposed to contaminant metabolites that, at times, are more bioactive than their precursors (e.g., oxychlordane, p,p'-DDE, OH- and MeSO$_2$-PCBs, and OH-PBDEs). Concentrations of legacy OCs (PCBs, DDTs, CHLs, CBzs, dieldrin, PCDD/Fs, and mirex) in tissues, blood, and eggs of Svalbard glaucous gulls have displayed the highest contamination levels among glaucous gull populations that inhabit Greenland (Cleemann et al. 2000), Jan Mayen (Gabrielsen et al. 1997), Alaska (Vander Pol et al. 2009), and the Canadian Arctic (Braune et al. 2005). To date, measurements obtained on more novel organohalogens (e.g., OH- and MeSO$_2$-containing metabolites, BFRs, and PFASs) in Svalbard glaucous gull samples generally confirm the spatial and trophodynamic trends of the legacy OC concentrations, whereas no clear trend emerges from surveys of trace elements and organometals. Using the glaucous gull as biosentinel species provides clear evidence that Svalbard and the European Arctic environment is exposed to a complex mixture of legacy and more recently introduced PBT-like substances.

Temporal trend analyses of legacy OC concentrations in the blood of breeding Svalbard glaucous gulls, collected between 1997 and 2006, revealed significant declines for PCBs, oxychlordane, and HCB. Nevertheless, projections based on this 9-yr temporal trend suggest the concentrations of legacy OCs in Svalbard glaucous gull samples will only decrease slowly in the decades to come. This is in general accord with time trend estimates for legacy OC levels in marine wildlife species from other arctic regions (Braune et al. 2005). However, one additional concern for the glaucous gull breeding in Svalbard is the relatively high blood and tissue concentrations of BFRs (e.g., PBDEs and HBCD), PFASs (e.g., PFSs and PFCAs), and cyclic siloxanes (e.g., siloxane-D5). Some of these emerging contaminants are known or suggested PBT candidates that have been shown to increase temporally in biota samples from many parts of the world, including the Arctic (Muir and Howard 2006).

The limited knowledge about and/or inability to control physiological condition parameters that influence the fate and toxicokinetics of organohalogens has been the foremost confounding aspect in contaminants research with the Svalbard glaucous gull and other wild bird species. Although the effect of age on organohalogen variation was found to be of minor importance in breeding Svalbard glaucous gulls, organohalogens were reported to vary among individuals as a function of the gender (maternal transfer), reproductive status (breeding vs. non-breeding), food item specialization (feeding ecology), and whole body composition (lipid and protein content and tissue selectivity).

Based on available threshold effect concentrations in avian species, certain contaminant classes reach tissue, blood, and/or egg concentration levels that may elicit adverse health effects in some of the most contaminated Svalbard glaucous gulls. However, because of data shortage in threshold effect levels in birds in general, and the debated validity of this approach due to differences in chemical sensitivity across animal species (Letcher et al. 2009), the biomarker approach was preferred, in Svalbard glaucous gull research, to assess the potential health impacts associated with contaminant exposure. These biological and ecological responses and effects cover most of the organizational levels of the biological systems from the molecular to the population level and thus provide a comprehensive health status assessment of breeding Svalbard glaucous gulls. More specifically, responses or effects have been reported on hepatic enzyme activity, retinoid and hormone homeostasis, basal metabolism and thermoregulation, immunity, genetic regulation, egg characteristics, feather growth, reproduction, behavior, as well as survival (Table 1). In some cases, the effects observed in adult gulls (and their chicks) were defined as adverse, within the natural response variation, and were suggested to be associated with reversible or irreversible changes in their organism. Because a remarkably larger number of correlations between biomarkers and contaminant concentrations have been reported in breeding males (Table 1), it can be postulated that organohalogen exposure-related effects in this top avian predator are gender-specific. This may indicate that male glaucous gulls during the breeding period are more sensitive to chemically induced changes mediated by contaminant exposure relative to females. Alternatively, this can be the result of the generally higher body burden of predominantly lipophilic organohalogens in males compared to females sampled shortly following the egg-laying period and up until hatching.

In conclusion, based on current understanding (weight-of-evidence) of biological and ecological response/effect studies, it is suggested that the general health of Svalbard glaucous gulls, mainly the males, is affected by the existing high concentrations of a complex array of PBT and PBT-like chemicals. The toxicological responses or effects, observed in various bird species exposed to these chemicals under laboratory settings, support this conclusion. However, the majority of glaucous gull studies have used a correlative approach, thus hampering the identification of potential causative agents and mechanisms of toxicity (Bustnes 2006). Therefore, despite a large and growing body of evidence on potential adverse health impacts in Svalbard glaucous gulls, no direct (causative) link can be established between the contaminant exposure and the marked population decline documented in Svalbard (Bear Island) since 1986 (Strøm 2007). In fact, the physiological stress potentially imposed by other factors or agents is either unstudied or poorly documented in this population. Such stressors may include pathogens, food scarcity, predator pressure, exposure to uncharacterized xenobiotic substances, and climate change. Nonetheless, organohalogens and other bioaccumulative xenobiotics deserve to be assessed for their potential to cause physiological stress in Svalbard glaucous gulls, in the context of other anthropogenic or natural stressors.

Appendix: Chemical acronym definition, sample size, year of collection, and the list of congeners and compounds composing the concentration sums shown in Fig. 2

Chemicals	Sample size	Year of collection	Congeners/compounds included in sums (Σ)
Σ_{41}Polychlorinated biphenyl (Σ_{41}PCB)	$n = 45$	2002	PCB-28, -31, -42, -44, -49, -52, -60, -64, -66/95, -70, -74, -97, -99, -101, -105, -110, -118, -128, -129/178, -138, -141, -146, -149, -151, -153, -158, -170/190, -171/202/156, -172, -174, -177, -179, -182/187, -180, -183, -194, -195, -200, -201, -203 and -206
Σ_3Dichlorodiphenyldichloroethane (Σ_3DDT)	$n = 45$	2002	p,p'-DDT, p,p'-DDD and p,p'-DDE
Σ_6Chlordane (Σ_6CHL)	$n = 45$	2002	oxychlordane, *trans*-chlordane, *cis*-chlordane, *trans*-nonachlor, *cis*-nonachlor and heptachlor epoxide

(continued)

Chemicals	Sample size	Year of collection	Congeners/compounds included in sums (Σ)
Σ_{21}Chlorobornane (Σ_{21}CHB)	$n = 10$	2002	B6-923a, B7-515, B7-1001, B7-1059a, B7-1450, B7-1474/B7-1440, B8-531, B8-789, B8-806, B8-810, B8-1412, B8-1413, B8-1414/B8-1945, B8-1471, B8-2229, B9-715, B9-718, B9-743/B9-2006, B9-1025, B9-1046 and B9-1679
Hexachlorobenzene (HCB)	$n = 45$	2002	
Dieldrin (DIEL)	$n = 45$	2002	
Σ_3Hexachlorocyclohexane (Σ_3HCH)	$n = 45$	2002	α-, β- and γ-HCH
Σ_{20}Polychlorinated naphthalene (Σ_{20}PCN)	$n = 10$	2004	PCN-28/43, -32, -33/34/37, -35, -47, -52/60, -53, -57, -58, -59, -61, -62, -63, -64/68, -65, -66/67, -69, -71/72, -73 and -74
Σ_2Mirex (Σ_2MIR)	$n = 45$	2002	mirex and *photo*-mirex
Σ_{13}Hydroxylated (OH)-PCB (Σ_{13}OH-PCB)	$n = 40$	2002	3'-OH-PCB-85, 4'-OH-PCB-104, 4-OH-PCB-107, 4-OH-PCB-112, 4'-OH-PCB-120, 4'-OH-PCB-130, 3'-OH-PCB-138, 4-OH-PCB-146, 4'-OH-PCB-159, 4-OH-PCB-165, 3'-OH-PCB-180, 4-OH-PCB-187 and 4-OH-PCB-193
Σ_{17}Methylsulfone (MeSO$_2$)-PCB (Σ_{17}MeSO$_2$-PCB)	$n = 40$	2002	3'-MeSO$_2$-PCB-49, 4'-MeSO$_2$-PCB-49, 3-MeSO$_2$-PCB-52, 4-MeSO$_2$-PCB-52, 4-MeSO$_2$-PCB-64, 3-MeSO$_2$-PCB-70, 4-MeSO$_2$-PCB-70, 4'-MeSO$_2$-PCB-87, 3'-MeSO$_2$-PCB-101, 4'-MeSO$_2$-PCB-101, 3-MeSO$_2$-PCB-110, 4-MeSO$_2$-PCB-110, 3'-MeSO$_2$-PCB-132, 4'-MeSO$_2$-PCB-132, 3-MeSO$_2$-PCB-149, unknown MeSO$_2$-Cl$_6$-PCB and 4-MeSO$_2$-PCB-174
3-MeSO$_2$-*p,p'*-DDE	$n = 40$	2002	

(continued)

Chemicals	Sample size	Year of collection	Congeners/compounds included in sums (Σ)
α-Hexabromocyclododecane (α-HBCD)	$n = 12$	2004	
Σ_{13}Polybrominated diphenyl ether (Σ_{13}PBDE)	$n = 12$	2004	PBDE-17, -28, -47, -66, -85, -99, -100, -138, -153, -154, -183, -190 and -209
Σ_{13}Methoxylated (MeO)-PBDE (Σ_{13}MeO-PBDE)	$n = 12$	2004	6'-MeO-PBDE-17, 4'-MeO-PBDE-17, 2'-MeO-PBDE-28, 4-MeO-PBDE-42, 6-MeO-PBDE-47, 3-MeO-PBDE-47, 5-MeO-PBDE-47/4'-MeO-PBDE-49, 6'-MeO-PBDE-49, 2'-MeO-PBDE-68, 6-MeO-PBDE-85, 6-MeO-PBDE-90/6-MeO-PBDE-99, 2-MeO-PBDE-123 and 6-MeO-PBDE-137
Σ_{14}OH-PBDE	$n = 12$	2004	4'-OH-PBDE-17, 6'-OH-PBDE-17, 4-OH-PBDE-42, 6-OH-PBDE-47, 3-OH-PBDE-47, 5-OH-PBDE-47, 4'-OH-PBDE-49, 6'-OH-PBDE-49, 2'-OH-PBDE-68, 6-OH-PBDE-85, 6-OH-PBDE-90, 6-OH-PBDE-99, 2-OH-PBDE-123 and 6-OH-PBDE-137
Σ_3 Perfluorosulfonate (Σ_3PFS)	$n = 10$	2004	perfluorobutane sulfonate, perfluorohexane sulfonate and perfluorooctane sulfonate
Σ_{10} Perfluorocarboxylate (Σ_{10}PFCA)	$n = 10$	2004	perfluoropentanoic acid, perfluorohexanoic acid, perfluorooctanoic acid, perfluorononanoic acid perfluorodecanoic acid, perfluoroundecanoic acid, perfluorododecanoic acid, perfluorotridecanoic acid, perfluorotetradecanoic acid and perfluoropentadecanoic acid

References

Berge JA, Brevik EM, Bjørge A, Følsvik N, Gabrielsen GW, Wolkers H (2004) Organotins in marine mammals and seabirds from Norwegian territory. J Environ Monit 6:108–112.

Bourne WRP, Bogan JA (1972) Polychlorinated biphenyls in North Atlantic seabirds. Mar Pollut Bull 3:171–175.

Braune BM, Outridge PM, Fisk AT, Muir DC, Helm PA, Hobbs K, Hoekstra PF, Kuzyk ZA, Kwan M, Letcher RJ, Lockhart WL, Norstrom RJ, Stern GA, Stirling I (2005) Persistent organic pollutants and mercury in marine biota of the Canadian Arctic: an overview of spatial and temporal trends. Sci Total Environ 351–352:4–56.

Breuner CW, Patterson SH, Hahn TP (2008) In search of relationships between the acute adrenocortical response and fitness. Gen Comp Endocrinol 157:288–295.

Buntin JD (1996) Neural and hormonal regulation of parental behaviors in birds. Adv Stud Behav 25:161–213.

Bustnes JO, Erikstad KE, Bakken V, Mehlum F, Skaare JU (2000) Feeding ecology and the concentration of organochlorines (OCs) in glaucous gulls. Ecotoxicol 9:179–186.

Bustnes JO, Skaare JU, Erikstad KE, Bakken V, Mehlum F (2001a) Whole blood concentrations of organochlorines as a dose metric for studies of the glaucous gull (*Larus hyperboreus*). Environ Toxicol Chem 20:1046–1052.

Bustnes JO, Bakken V, Erikstad KE, Mehlum F, Skaare JU (2001b) Patterns of incubation and nest site attentiveness in relation to organochlorine (PCB) contamination in glaucous gulls. J Appl Ecol 38:791–801.

Bustnes JO, Folstad I, Erikstad KE, Fjeld M, Miland ØO, Skaare JU (2002) Blood concentration of organochlorines and wing feather asymmetry in glaucous gull. Funct Ecol 16:617–622.

Bustnes JO, Bakken V, Skaare JU, Erikstad KE (2003a) Age and accumulation of persistent organochlorines: a study of arctic breeding glaucous gulls. Environ Toxicol Chem 22:2173–2179.

Bustnes JO, Erikstad KE, Skaare JU, Bakken V, Mehlum F (2003b) Ecological effects of organochlorine pollutants in the Arctic: a study of the glaucous gull. Ecol Applic 13:504–515.

Bustnes JO, Hanssen SA, Folstad I, Erikstad KE, Hasselquist D, Skaare JU (2004) Immune function and organochlorine pollutants in arctic breeding glaucous gull. Arch Environ Contam Toxicol 47:530–541.

Bustnes JO, Miland ØO, Fjeld M, Erikstad KE (2005) Relationships between ecological variables and four organochlorine pollutants in an arctic glaucous gull (*Larus hyperboreus*) population. Environ Pollut 136:175–185.

Bustnes JO (2006) Pinpointing potential causative agents in mixtures of persistent organic pollutants in observational field studies: a review of glaucous gull studies. J Toxicol Environ Health A 69:97–108.

Bustnes JO, Erikstad KE, Hanssen SA, Folstad I, Skaare JU (2006) Anti-parasite treatment removes negative effects of environmental pollutants on reproduction in an arctic seabird. Proc R Soc B 273:3117–3122.

Bustnes JO, Gabrielsen GW, Verreault J (Submitted) Temporal trends (1997–2006) of persistent organic pollutants in relation to biological and climatic factors: a study of glaucous gulls from the European Arctic. Environ Sci Technol.

Chastel O, Lacroix A, Kersten M (2003) Pre-breeding energy requirements: thyroid hormone, metabolism and the timing of reproduction in house sparrows *Passer domesticus*. J Avian Biol 34:298–306.

Clarke GM (1995) Relationship between developmental stability and fitness: application for conservation biology. Conserv Biol 9:18–24.

Cleemann M, Riget F, Paulsen GB, Dietz R (2000) Organochlorines in Greenland glaucous gulls (*Larus hyperboreus*) and Icelandic gulls (*Larus glaucoides*). Sci Total Environ 245:117–130.

Daelemans FF, Mehlum F, Schepens PJ (1992) Polychlorinated biphenyls in two species of Arctic seabirds from the Svalbard area. Bull Environ Contam Toxicol 48:828–834.

de Wit CA, Fisk AT, Hobbs KE, Muir DCG, Gabrielsen GW, Kallenborn R, Krahn MM, Norstrom RJ, Skaare JU (2004) AMAP Assessment 2002: Persistent organic pollutants in the Arctic. Arctic Monitoring and Assessment Programme (AMAP), Oslo, Norway, xvi + 310 pp.

de Wit CA, Alaee M, Muir DCG (2006) Levels and trends of brominated flame retardants in the Arctic. Chemosphere 64:209–233.

Dubois M, Pfohl-Leszkowicz A, Grosse Y, Kremers P (1995) DNA adducts and P450 induction in human, rat and avian liver cells after exposure to polychlorobiphenyls. Mutat Res 345: 181–190.

Ellis HI, Gabrielsen GW (2002) Energetics of free-ranging seabirds. In: Schreiber EA, Burger J (eds) Biology of marine birds. CRC Press, Washington, DC, pp 359–407.

Erikstad KE, Moum T, Bustnes JO, Reiertsen TK (2009) High levels of organochlorines have detrimental effects on the sex allocation strategies in arctic glaucous gulls. Funct Ecol.

Fox GA, Jeffrey DA, Williams KS, Kennedy SW, Grasman KA (2007) Health of herring gulls (*Larus argentatus*) in relation to breeding location in the early 1990s. I. Biochemical measures. J Toxicol Environ Health A 70:1443–1470.

Gabrielsen GW (2007) Levels and effects of persistent organic pollutants in arctic animals. In: Orbaek JB, Kallenborn R, Tombre I, Hegseth EN, Falk-Petersen S, Hoel AH (eds) Arctic-alpine ecosystems and people in a changing environment. Springer-Verlag, Berlin, Germany, pp 377–412.

Gabrielsen GW, Skaare JU, Polder A, Bakken V (1995) Chlorinated hydrocarbons in glaucous gulls (*Larus hyperboreus*) in the southern part of Svalbard. Sci Total Environ 160/161: 337–346.

Gabrielsen GW, Alsos IG, Brekke B (1997) Undersøkelse av jord, fisk og sjøfugl i området rundt avfallsfyllingen på Jan Mayen. Norsk Polarinstutts Rapportserie 104:31. [in Norwegian]

Gaston AJ, Descamps S, Gilchrist HG (2009) Reproduction and survival of glaucous gulls breeding in an Arctic seabird colony. J Field Ornithol 80:135–145.

Grasman KA (2002) Assessing immunological function in toxicological studies of avian wildlife. Integr Comp Biol 42:34–42.

Hakk H, Letcher RJ (2003) Metabolism in the toxicokinetics and fate of brominated flame retardants – a review. Environ Int 29:801–828.

Haukås M, Berger U, Hop H, Gulliksen B, Gabrielsen GW (2007) Bioaccumulation of per- and polyfluorinated alkyl substances (PFAS) in selected species from the Barents Sea food web. Environ Pollut 148:360–371.

Henriksen EO, Gabrielsen GW, Skaare JU (1998a) Validation of the use of blood samples to assess tissue concentrations of organochlorines in glaucous gulls. Chemosphere 37:2627–2643.

Henriksen EO, Brunstrøm B, Skaare JU, Gabrielsen GW (1998b) Bioassay-derived TCDD equivalents and *mono-ortho* PCB concentrations in liver of glaucous gulls, *Larus hyperboreus*, from Svalbard. Organohalogen Comp 39:415–418.

Henriksen EO, Gabrielsen GW, Skaare JU, Skjegstad N, Jenssen BM (1998c) Relationships between PCB levels, hepatic EROD activity and plasma retinol in glaucous gulls, *Larus hyperboreus*. Mar Environ Res 46:45–49.

Henriksen EO, Gabrielsen GW, Trudeau S, Wolkers J, Sagerup K, Skaare JU (2000) Organochlorines and possible biochemical effects in glaucous gulls (*Larus hyperboreus*) from Bjørnøya, the Barents Sea. Arch Environ Contam Toxicol 38:234–243.

Herzke D, Gabrielsen GW, Evenset A, Burkow IC (2003) Polychlorinated camphenes (toxaphenes), polybrominated diphenylethers and other halogenated organic pollutants in glaucous gull (*Larus hyperboreus*) from Svalbard and Bjørnøya (Bear Island). Environ Pollut 121:293–300.

Hop H, Borgå K, Gabrielsen GW, Kleivane L, Skaare JU (2002) Food web magnification of persistent organic pollutants in poikilotherms and homeotherms. Environ Sci Technol 36:2589–2597.

Hose JE, Guillette LJ (1995) Defining the role of pollutants in the disruption of reproduction in wildlife. Environ Health Perspect 103:87–91.

Ims RA, Fuglei E (2005) Trophic interaction cycles in tundra ecosystems and the impact of climate change. BioScience 55:311–322.

Jæger I, Hop H, Gabrielsen GW (2009) Biomagnification of mercury in selected species from an Arctic marine food web in Svalbard. Sci Total Environ 407:4744–4751.

Jenssen BM (2006) Endocrine-disrupting chemicals and climate change: a worst-case combination for Arctic marine mammals and seabirds? Environ Health Perspect 114:76–80.

Knudsen LB, Gabrielsen GW, Verreault J, Barrett R, Skåre JU, Polder A, Lie E (2005) Temporal trends of brominated flame retardants, cyclododeca-1,5,9-triene and mercury in eggs in four seabird species from Northern Norway and Svalbard. Report no. SFT 942/2005. Norwegian Pollution Control Authority (SFT), Oslo, Norway, 44 pp.

Knudsen LB, Polder A, Føreid S, Lie E, Gabrielsen GW, Barrett R, Skåre JU (2006) Nona- and deca-brominated diphenylethers in seabird eggs from Northern Norway and Svalbard. Report no. SFT 952/2006. Norwegian Pollution Control Authority (SFT), Oslo, Norway, 30 pp.

Knudsen LB, Sagerup K, Polder A, Schlabach M, Josefsen TD, Strøm H, Skåre JU, Gabrielsen GW (2007) Halogenated organic contaminants (HOCs) and mercury in dead or dying seabirds on Bjørnøya (Svalbard). Report no. SFT 977/2007. Norwegian Pollution Control Authority (SFT), Oslo, Norway, 45 pp.

Krøkje A, Bingham C, Tuven RH, Gabrielsen GW (2006) Chromosome aberrations and DNA strand breaks in glaucous gull (*Larus hyperboreus*) chicks fed environmentally contaminated gull eggs. J Toxicol Environ Health A 69:159–174.

Letcher RJ, Klasson-Wehler E, Bergman Å (2000) Methyl sulfone and hydroxylated metabolites of polychlorinated biphenyls. In: Paasivita J (ed) The handbook of environmental chemistry–new types of persistent halogenated compounds, vol. 3, Part K. Springer-Verlag, Heidelberg, Germany, pp 315–359.

Letcher RJ, Bustnes JO, Dietz R, Jenssen BM, Jørgensen EH, Sonne C, Verreault J, Vijayan MM, Gabrielsen GW (2009) Exposure and effects assessment of persistent organohalogen contaminants in Arctic wildlife and fish. Sci Total Environ.

McKinney MA, Peacock E, Letcher RJ (2009) Sea ice-associated diet change increases the levels of chlorinated and brominated contaminants in polar bears. Environ Sci Technol 43:4334–4339.

McNabb FMA (2005) Biomarkers for the assessment of avian thyroid disruption by chemical contaminants. Avian Poult Biol Rev 16:3–10.

Moriarty F, Bell AA, Hanson H (1986) Does p,p'-DDE thin eggshells Environ Pollut Series A-Ecolo biol 40:257–286.

Muir DC, Howard PH (2006) Are there other persistent organic pollutants? A challenge for environmental chemists. Environ Sci Technol 40:7157–7166.

Norheim G (1987) Levels and interactions of heavy metals in seabirds from Svalbard and the Antarctic. Environ Pollut 47:83–94.

Norheim G, Kjos-Hanssen B (1984) Persistent chlorinated hydrocarbons and mercury in birds caught off the west-coast of Spitsbergen. Environ Pollut Series A-Ecolo Biol 33:43–152.

Østby L, Gabrielsen GW, Krøkje A (2005) Cytochrome P4501A induction and DNA adduct formation in glaucous gulls (*Larus hyperboreus*), fed with environmentally contaminated gull eggs. Ecotoxicol Environ Saf 62:363–375.

Peakall DB (1996) Disrupted patterns of behavior in natural populations as an index of ecotoxicity. Environ Health Perspect 104:331–335.

Pusch K, Schlabach M, Prinzinger R, Gabrielsen GW (2005) Gull eggs – food of high organic pollutant content? J Environ Monit 7:635–639.

Rolland RM (2000) A review of chemically-induced alterations in thyroid and vitamin A status from field studies of wildlife and fish. J Wildlife Dis 36:615–635.

Ross MS, Verreault J, Letcher RJ, Gabrielsen GW, Wong CS (2008) Chiral organochlorine contaminants in the blood and eggs of glaucous gulls (*Larus hyperboreus*) from the Norwegian Arctic. Environ Sci Technol 42:7181–7186.

Sagerup K, Henriksen EO, Skorping A, Skaare JU, Gabrielsen GW (2000) Intensity of parasitic nematodes increases with organochlorine levels in the glaucous gull. J Appl Ecol 37:532–539.

Sagerup K, Henriksen EO, Skaare JU, Gabrielsen GW (2002) Intraspecific variation in trophic feeding levels and organochlorine concentrations in glaucous gulls (*Larus hyperboreus*) from Bjørnøya, the Barents Sea. Ecotoxicol 11:119–125.

Sagerup K, Helgason LB, Polder A, Strøm H, Josefsen TD, Skaare JU, Gabrielsen GW (2009a) Persistent organic pollutants and mercury in dead and dying glaucous gulls (*Larus hyperboreus*) at Bjørnøya (Svalbard). Sci Total Environ.

Sagerup K, Savinov V, Savinova T, Kuklin VV, Muir DCG, Gabrielsen GW (2009b) Persistent organic pollutants, heavy metals and parasites in the glaucous gull (*Larus hyperboreus*) on Spitsbergen. Environ Pollut 157:2282–2290.

Sagerup K, Larsen HJS, Skaare JU, Johansen GM, Gabrielsen GW (2009c) The toxic effects of multiple persistent organic pollutant exposures on the post-hatch immunity maturation of glaucous gulls. J Toxicol Environ Health A 72:870–883.

Sanderson T, van den Berg M (2003) Interactions of xenobiotics with the steroid hormone biosynthesis pathway. Pure Appl Chem 75:1957–1971.

Savinov VM, Gabrielsen GW, Savinova TN (2003) Cadmium, zinc, copper, arsenic, selenium and mercury in seabirds from the Barents Sea: levels, inter-specific and geographical differences. Sci Total Environ 306:133–158.

Savinova TN, Polder A, Gabrielsen GW, Skaare JU (1995) Chlorinated hydrocarbons in seabirds from the Barents Sea area. Sci Total Environ 160/161:497–504.

Strøm H (2007) Distribution of seabirds on Bjørnøya. In: Anker-Nilssen T, Barrett RT, Bustnes JO, Erikstad KE, Fauchald P, Lorentsen S-H, Steen H, Strøm H, Systad GH, Tveraa T (eds) SEAPOP studies in the Lofoten and Barents sea area in 2006. Report no. 249. Norwegian Institute for Nature Research (NINA), Tromsø, Norway, 63 pp.

Sures B (2006) How parasitism and pollution affect the physiological homeostasis of aquatic hosts. J Helminthol 80:151–157.

Teuten EL, Xu L, Reddy CM (2005) Two abundant bioaccumulated halogenated compounds are natural products. Science 307:917–920.

Trivers RL, Willard DE (1973) Natural selection of parental ability to vary the sex ratio of offspring. Science 179:90–92.

Ucán-Marín F, Arukwe A, Mortensen A, Gabrielsen GW, Fox GA, Letcher RJ (2009) Recombinant transthyretin purification and competitive binding with organohalogen compounds in two gull species (*Larus argentatus* and *Larus hyperboreus*). Toxicol Sci 107:440–450.

Vander Pol SS, Becker PR, Ellisor MB, Moors AJ, Pugh RS, Roseneau DG (2009) Monitoring organic contaminants in eggs of glaucous and glaucous-winged gulls (*Larus hyperboreus* and *Larus glaucescens*) from Alaska. Environ Pollut 157:755–762.

Verboven N, Verreault J, Letcher RJ, Gabrielsen GW, Evans E (2008) Maternally derived testosterone and 17b-estradiol in the eggs of Arctic-breeding glaucous gulls in relation to persistent organic pollutants. Comp Biochem Physiol C: Toxicol Pharmacol 148: 143–151.

Verboven N, Verreault J, Letcher RJ, Gabrielsen GW, Evans NP (2009a) Adrenocortical function of Arctic-breeding glaucous gulls in relation to persistent organic pollutants. Gen Comp Endocrinol.

Verboven N, Verreault J, Letcher RJ, Gabrielsen GW, Evans E (2009b) Nest temperature and parental behaviour of Arctic-breeding glaucous gulls exposed to persistent organic pollutants. Anim Behav 77:411–418.

Verboven N, Verreault J, Letcher RJ, Gabrielsen GW, Evans E (2009c) Differential investment in eggs by Arctic-breeding glaucous gulls (*Larus hyperboreus*) exposed to persistent organic pollutants. Auk 126:123–133.

Verreault J, Skaare JU, Jenssen BM, Gabrielsen GW (2004) Effects of organochlorine contaminants on thyroid hormone levels in Arctic breeding glaucous gulls, *Larus hyperboreus*. Environ Health Perspect 112:532–537.

Verreault J, Letcher RJ, Muir DCG, Chu S, Gebbink WA, Gabrielsen GW (2005a) New organochlorine contaminants and metabolites in plasma and eggs of glaucous gulls (*Larus hyperboreus*) from the Norwegian Arctic. Environ Toxicol Chem 24:2486–2499.

Verreault J, Gabrielsen GW, Chu S, Muir DCG, Andersen M, Hamaed A, Letcher RJ (2005b) Flame retardants and methoxylated and hydroxylated polybrominated diphenyl ethers in two Norwegian Arctic top predators: Glaucous gulls and polar bears. Environ Sci Technol 39: 6021–6028.

Verreault J, Houde M, Gabrielsen GW, Berger U, Haukås M, Letcher RJ, Muir DCG (2005c) Perfluorinated alkyl substances in plasma, liver, brain and eggs of glaucous gulls (*Larus hyperboreus*) from the Norwegian Arctic. Environ Sci Technol 39:7439–7445.

Verreault J, Agudo Villa R, Gabrielsen GW, Skaare JU, Letcher RJ (2006a) Maternal transfer of organohalogen contaminants and metabolites to eggs of Arctic-breeding glaucous gulls. Environ Pollut 144:1053–1060.

Verreault J, Letcher RJ, Ropstad E, Dahl E, Gabrielsen GW (2006b) Organohalogen contaminants and reproductive hormones in incubating glaucous gulls (*Larus hyperboreus*) from the Norwegian Arctic. Environ Toxicol Chem 25:2990–2996.

Verreault J, Gebbink WA, Gauthier LT, Gabrielsen GW, Letcher RJ (2007a) Brominated flame retardants in glaucous gulls from the Norwegian Arctic: more than just an issue of polybrominated diphenyl ethers. Environ Sci Technol 41:4925–4931.

Verreault J, Shahmiri S, Gabrielsen GW, Letcher RJ (2007b) Organohalogen and metabolically-derived contaminants and associations with whole body constituents in Norwegian Arctic glaucous gulls. Environ Int 33:823–830.

Verreault J, Bech C, Letcher RJ, Ropstad E, Dahl E, Gabrielsen GW (2007c) Organohalogen contamination in breeding glaucous gulls from the Norwegian Arctic: Associations with basal metabolism and circulating thyroid hormones. Environ Pollut 145:138–145.

Verreault J, Verboven N, Gabrielsen GW, Letcher RJ, Chastel O (2008) Changes in prolactin in a highly organohalogen-contaminated Arctic top predator seabird, the glaucous gull. Gen Comp Endocrinol 156:569–576.

Warner NA, Norstrom RJ, Wong CS, Fisk AT (2005) Enantiomeric fractions of chiral polychlorinated biphenyls provide insights on biotransformation capacity of arctic biota. Environ Toxicol Chem 24:2763–2767.

Fenamiphos and Related Organophosphorus Pesticides: Environmental Fate and Toxicology

Tanya Cáceres, Mallavarapu Megharaj, Kadiyala Venkateswarlu, Nambrattil Sethunathan, and Ravi Naidu

Contents

1	Introduction	118
	1.1 Chemistry of Organophosphorus Pesticides	118
	1.2 Mode of Action of OP Compounds	119
	1.3 Metabolism of OP Compounds	121
2	Fate and Transport of OP Compounds in the Environment	121
	2.1 Transport Processes	122
	2.2 Chemical Processes	126
	2.3 Biological Processes	127
3	Soil Bioavailability of OP Compounds	129
4	OP Pesticides and Health	132
5	Fenamiphos	133
	5.1 Environmental Fate	134
	5.2 Ecotoxicology	139
	5.3 Toxicity and Fate in Mammals	142
6	Isofenphos	144
	6.1 Environmental Fate	145
	6.2 Ecotoxicology	146
	6.3 Toxicity in Mammals	147
	6.4 Fate in Humans and Animals	149
7	Coumaphos	150
	7.1 Environmental Fate	151
	7.2 Ecotoxicology	152
8	Future Research	153
9	Summary	153
	References	154

M. Megharaj (✉)
Centre for Environmental Risk Assessment and Remediation (CERAR), Cooperative Research Centre for Contamination Assessment and Remediation of the Environment (CRC CARE), University of South Australia, Mawson Lakes, 5095, SA, Australia
e-mail: megharaj.mallavarapu@unisa.edu.au

1 Introduction

Organophosphorus (OP) compounds are among the most common chemical classes used in crop and livestock protection and account for an estimated 34% of worldwide insecticide sales (Singh and Walker 2006). During the last 60 yr, approximately 150 different OP chemicals have been used to protect crops, livestock, and human health (Casida and Quistad 2004). In recent years, the OP compounds have been the most widely used group of insecticides in Australia. In addition to fenamiphos (nematicide), isofenphos, and coumaphos (insecticides), the most commonly used OP pesticides include parathion methyl, chlorpyrifos, dimethoate, and profenfos. Approximately 500 t of OP-active ingredients comprising about 30 distinct compounds have been used annually in Australia for many years (Australian Academy of Technological Sciences and Engineering 2002).

Because some members of the OP pesticide group are very toxic to aquatic invertebrates, there is serious concern about the environmental fate of these chemicals. Many reports have been published on water and food contamination, and the consequential effects on wildlife and health, from exposure to OP pesticides such as fenamiphos and isofenphos (Franzmann et al. 2000; Patrick et al. 2001; Pazy Miño et al. 2002; Pesticide Residues Committee (UK) 2007). Moreover, the use of these compounds results in the poisoning of about 200,000 people per year worldwide (Kalvaci et al. 2009). To avoid exposure incidents from use of highly toxic products, the residential use of organophosphates such as chlorpyrifos and diazinon has been reduced in the USA (Stone et al. 2009). Guiatart et al. (2009) studied the poisoning of European wildlife by different xenobiotics, including OP pesticides, over the last 10 yr. The cholinesterase inhibitors were responsible for 42.5% of all wild bird fatalities in France. In Greece, organophosphates like methamidophos, parathion, methyl phorate, and fenthion were reported to be associated with 50% of all cases of wild bird deaths. Other organophosphates, such as malathion and fenthion that are currently used on olive trees and vines, were detected in the gastric content of three raptors. In addition, organophosphates were confirmed to be involved in the poisoning of wild mammals in northern Greece. Such reports are important because they disclose the behavior and potential harm caused by these chemicals in the environment, and these reports are useful in identifying data gaps for remediation by future research.

In this review, we emphasize (i) the fate of OP insecticides in the environment, in general, and (ii) the environmental fate and toxicity of fenamiphos and related compounds such as isofenphos and coumaphos, in particular.

1.1 Chemistry of Organophosphorus Pesticides

Organophosphates have the following general structure:

$$\begin{array}{l} X \\ R\text{-}O\text{-}P\text{-}O\text{-}Ar \\ R \end{array}$$

where letters represent: R is a methyl or ethyl group, X is sulfur (S), O is oxygen, and Ar a more complex moiety, which is often aromatic. The degradation of these compounds, usually at R, X, and /or Ar groups, results in the breakup of the molecule (Ragnarsdottir 2000). OP compounds are a diverse family of chemicals. The nomenclature of these compounds is complex and their classification may follow various systems depending on the part of the structure used in the classification.

The majority of OPs may be identified as derivatives of phosphoric acid. Accordingly, there are 12 types of OP insecticides, and the partial structures of these compounds are depicted in Fig. 1. The original phosphate pesticides were triesters of phosphoric acid and may be considered as the standard model of the complete family of OP compounds. In the triesters, all four atoms surrounding the phosphorus are oxygen atoms. These compounds are highly reactive, have short residual activity, and usually can be applied to crops shortly before harvest. The phosphorothionates are abundant and retain sulfur in their structure, usually in a P=S moiety. Parathion, diazinon, and chlorpyrifos are examples of phosphorothionates. The phosphorothiolates also contain S in their structure. Members of this subclass of phosphates are generally more toxic and are usually applied as soil or plant systemic insecticides. An example of this group is demeton (II-isomer).

Another important group of OP pesticides is the phosphorthionothiolates; this group is characterized by having an S atom as a P=S moiety along with another S as a thioester. Examples of this subclass are the highly toxic phorate and terbufos. The phosphorodithioates are also phosphorthionothiolates; ethoprop and ebufos are examples. The phosphorous acids form amides as well as esters. OP amide derivatives include the phosphoramides and phosphoramidates, examples of which are fenamiphos, phosfolan, and mephospholan. Phosphoramidothionates such as propetamphos and isofenphos and phosphoramidothiolates like methamidophos and acephate belong to this subclass. The phosphonate and related OP compounds have one substituent, which is attached by a phosphorus–carbon bond. Examples of this group are trichlorfon (phosphonate) and fonofos (phosphonothionothiolate) (Chambers and Levi 1992).

1.2 Mode of Action of OP Compounds

OP pesticides are neurotoxins that inhibit the breakdown of the neurotransmitter acetylcholine (Chambers and Levi 1992; Ragnarsdottir 2000). Acetylcholine is an important substance linked to the transmission of nerve impulses in the brain, skeletal muscles, and other parts of the anatomy (Schmidt-Nielsen 1983). The OP compounds inactivate the enzyme acetyl cholinesterase (AChE); this enzyme has an important role in the functioning of nerve cells, because it catalyzes the rapid hydrolysis of acetylcholine (Raushel 2002).

The reaction mechanism for the enzymatic hydrolysis of acetylcholine is irreversible and is known as transesterification. Acetylcholine is the normal substrate of AChE. Specifically, a serine hydroxyl group at the catalytic center of the enzyme is acetylated, releasing the choline moiety. Subsequent hydrolysis of AChE completes

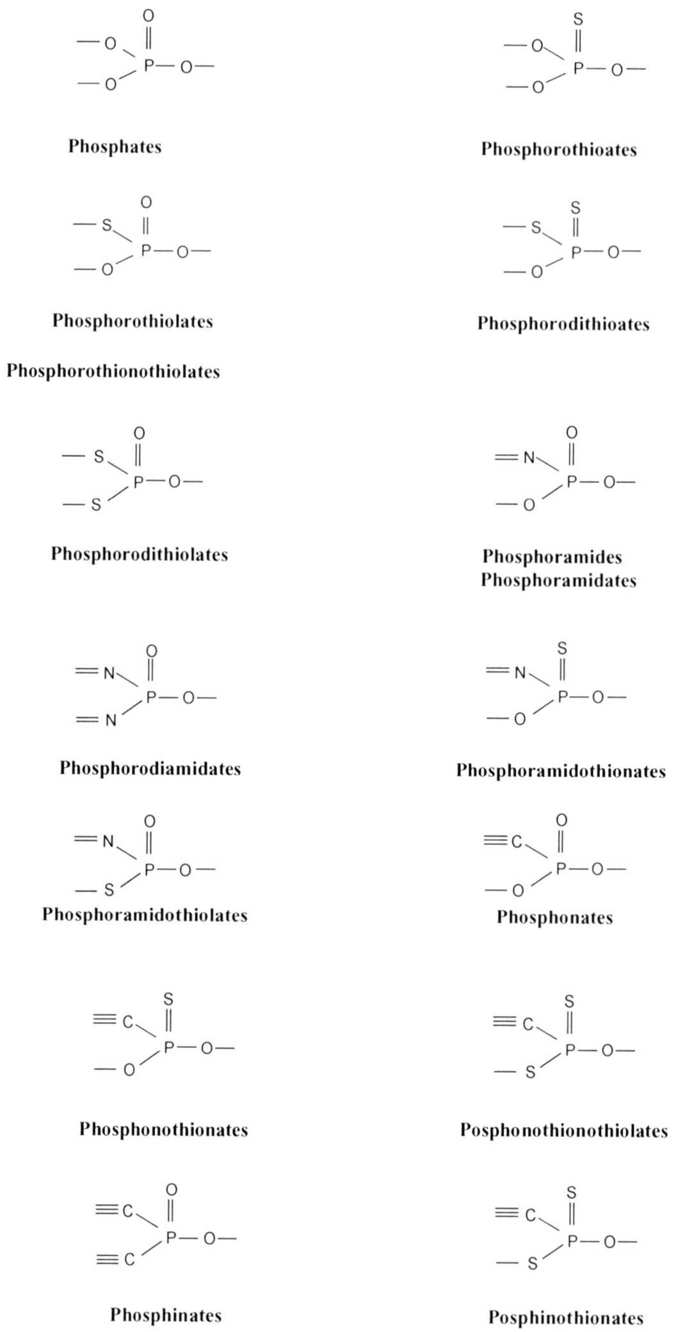

Fig. 1 The basic structural frames for the different organophosphorous (OP) esters

the reaction. OP molecules inhibit AChE and mimic its action by phosphorylating the serine hydroxyl moiety. Hydrolysis of the phosphoryl-AChE, however, is exceedingly slow and the catalytic activity of the enzyme molecule is lost (Chambers and Levi 1992).

1.3 Metabolism of OP Compounds

Cytochrome P450 is the terminal oxidase of the cytochrome P450-dependent monooxygenase system (P450) and is very important in the metabolism of OP compounds. For example, both desulfuration and dearylation of phosphorothionates are oxidative and P450 dependent. It has been proposed that both processes share a common intermediate, a phosphooxathiiran. This three-membered ring containing P, O, and S could be formed from P=S by a reaction parallel to epoxidation of a C=C moiety. Elimination of the S would yield the oxon. The dearylation is less obvious, because the final products are expected from hydrolysis alone. It is hypothesized that hydrolysis of the P–O–aryl bond occurs with the phosphooxathiiran, and O rather than S is eliminated from the acidic product. Interestingly, different P450 isozymes may produce different ratios of metabolic products (Chambers and Levi 1992). In addition to the cytochrome P450 monooxygenases, there are other flavin-containing monooxygenases (FMO) that also play an important role in the metabolism of xenobiotics (Levi and Hodson 1992).

The glutathione transferases are also involved in the metabolism of OPs. Reduced glutathione is known to be an acceptor for several alkyl and aryl transfer reactions. Methyl iodide and dichloronitrobenzene are commonly utilized as model substrates for glutathione transferase enzymes. Some OP compounds are also known to be substrates for glutathione-dependent metabolism. Approximately 25 OP compounds, e.g., methyl parathion and azinphos-methyl, have been reported to undergo significant biotransformation as a result of conjugation with glutathione (Sultatos 1992). There are two types of esterases which are important in detoxification of OP compounds. The A-esterases, e.g., paraoxonase, can actively hydrolyze some OP compounds, although their activity is limited. The B-esterases such as aliesterases (carboxylesterases) are believed to contribute to OP detoxification by acting as alternative phosphorylation sites. Finally, the aliesterases have been identified as active molecules in the degradation of OP compounds that contain carboxylic ester moieties within the leaving group. The best example of this degradation pattern is malathion which owes most of its low mammalian toxicity to rapid hydrolysis of carboxylester groups in the molecule by liver esterases (Chambers and Levi 1992).

2 Fate and Transport of OP Compounds in the Environment

The presence of OP compounds in soil, groundwater, surface water, and in the atmosphere results from the use of these compounds as pest control agents on crops and animals as well as disposal of cattle and sheep dips on farmland. OP compounds have been found in soil (Ou et al. 1993, 1994), river water (Kikuchi et al. 2000; Bondarenko et al. 2004; Anderson et al. 2006; Phillips et al. 2006), groundwater (Franzmann et al. 2000; Patrick et al. 2001), river sediments (Long et al. 1998), estuarine water and sediments (Ortega et al. 2006), and in the atmosphere (McConnell et al. 1997; Feigenbrugel et al. 2006). OP insecticides have been detected in needles of *Pinus ponderosa* in the Sierra Nevada Mountains away from the zone of current application in California (Aston and Seiber 1997). Although the environmental persistence of OP insecticides is known to be affected by transport processes and by chemical and biological degradation processes (Ragnarsdottir 2000), the appearance of OP residues in remote environments demonstrate that these pesticides can be transported long distances by air and water. Such detections also underscore the importance of knowing more of their fate and transport in the environment.

2.1 Transport Processes

2.1.1 Sorption

Sorption influences the behavior and fate of pesticides in the environment (Caceres et al. 2002). Sorption occurs when a pesticide molecule comes in contact with soil constituents and often establishes, within hours, a pseudo-equilibrium with these constituents. The term sorption is defined as the transfer of a solute between a fluid and a solid phase. This process requires consideration of both solute and solvent behavior. Sorption depends on three factors, which often interact in a complex manner: the molecular properties of the solute, soil constitution, and the experimental conditions under which the sorption is studied (Koskinen and Harper 1990). OP insecticides are known to sorb to mineral surfaces or organic matter present in soils. Adsorption of solutes such as OP substances to soil solids results from mineral surfaces having a pH-dependent surface charge in aqueous solutions (Ragnasdorttir 2000).

Since sorption takes place in an aqueous medium, interactions between water and solutes are of particular importance. For example, OPs that are less water soluble and, thereby, more hydrophobic are often more strongly sorbed to soil. The initial behavior of a pesticide in soil will be affected by the physical nature of the formulation and the presence of additives like wetting agents, emulsifiers, and surfactants. Unfortunately, there is little information available on the effects of formulation constituents on OP behavior in soil. Clays and iron oxides are among the most important soil mineral fractions that affect adsorption. This is both because of their abundance and their surface properties. In addition, adsorption on clays is conditioned by water content, acidity/basicity, and the ionic composition of the surfaces. Compensating cations can act in three ways:

- compete with positively charged organic molecules for adsorption sites (e.g., the herbicides paraquat and atrazine);
- act directly as adsorption sites through the formation of coordination bonds; or

- cations such as Al^{3+} and Fe^{3+} can form hydroxides on the clay surface. This results in a great increase of mineral adsorption capacity.

Soil organic matter is not yet so well characterized as to allow its relationship between structure and properties to be exactly defined. However, it has been observed that the adsorption of the non-ionic pesticides (e.g., the herbicides diuron and linuron) varies greatly according to the nature of the organic matter with which they come into contact. Soil organic matter is not a single material but a mixture of solid and semisolid materials that have a range of properties depending on the history of the sample and its age (Wauchope et al. 2002). Soil organic matter is the entity mainly responsible for the sorption of pesticides, especially non-ionic pesticides. It is becoming increasingly clear that not only the amount of organic carbon but also its chemistry has a major influence on pesticide sorption rates (Ahmad et al. 2003; Ahmad and Kookana 2002). Adsorption rates depend on several factors in addition to the nature of the adsorbent and pesticide. These factors include temperature, ionic composition of the solution, and soil–water ratios.

Sorption is assessed using direct and indirect measurements; both of these involve equilibration of a soil sample with an aqueous pesticide solution of known initial concentration. The direct method involves measurement of both solution concentration and sorbed concentration; after equilibration is reached, the sorbed pesticide is displaced from the soil by an appropriate solvent and directly quantified. The indirect method determines the quantity of sorbed pesticide by measuring the change in solution concentration resulting from soil sorption of pesticide from solution; in this case, the quantity of sorbed solute is assumed to equal the quantity lost from solution. Sorption can be described by using sorption isotherms. The most commonly used expression of a sorption isotherm is the Freundlich equation, which defines a nonlinear relationship between the amount sorbed and the equilibrium solution concentration:

$$S = K_f C^n$$

where S is the concentration of a pesticide sorbed by the soil (expressed as mg/kg), K_f is the Freundlich sorption coefficient (1/kg), C is the equilibrium solution concentration (mg/L), and n is a power function related to the sorption mechanism. When the value is unity, we have the simple linear isotherm:

$$S = K_d C$$

where K_d is the sorption coefficient (L/kg). Many studies show that the sorption of pesticides closely depends on the organic content in the soil and hence sorption coefficients are normalized based on the organic carbon content. $K_{oc} = K_d/\%$ organic carbon. The K_d values calculated for several OP pesticides, based on soil properties, are presented in Table 1.

Table 1 Calculated K_d soil adsorption values, leaching potential and hydrolysis kinetics for selected OP pesticides

Pesticide	K_d	Leaching potential	Hydrolysis (half-life, days)	pH of hydrolysis
Azinphos-methyl	8.94	Small		
Chlorethoxyfos	63.2			
Chlorpyrifos	610	Small	72	5
			72	7
			16	9
Chlorpyrifos oxon			85	5
			6	7
			<1	9
Coumaphos	61–298	Small		
Diazinon	12–35	Large	2	5
			23	7
			12	9
Dicrotophos	1.01	Medium		
Dimethoate	0.45	Medium		
Disulfoton	14.7	Small		
Ethoprophos	1.69	Large		
Fenamiphos	3.84	Large		
Fenthion	18.2	Small		
Fonofos		Medium		
Isofenphos	8.52	Medium		
Malathion	575	Small		
Metamidofos	0.089	Small		
Parathion	26		301	5
			168	7
			105	8
Phorate	6.47	Medium	4	7
Phorate sulfoxide			175	7
Phorate sulfone			56	7
Terbufos	20.7	Small		

Adapted from Weber et al. (2004), Racke (1992), National Pesticide Information Centre (1994), Kearney et al. (1986), Leyhe (2004), Valverde García et al. (1992), Cooke et al. (2004), Koleli et al. (2007), Wauchope et al. (2002), and Baskaran et al. (1996)

2.1.2 Volatilization

Volatilization is an important process in the dissipation of some OP compounds from the surface of moist soil, foliage, or water. The severity of this process is linked to several factors such as vapor pressure (vp), solubility, adsorptive behavior, and persistence of the compound, as well as environmental characteristics like temperature, moisture, and air movement (Racke 1992). The vp varies with OP compounds; for example, dichlorvos has a vp of 1.0×10^{-2} mmHg and is volatilized rapidly under environmental conditions; that is the reason for the high fumigant action of dichlorvos and the resulting effective insect control. In contrast, EPN (ethyl *p*-nitrophenyl thiobenzene) has a very low vp and is virtually non-volatile. The volatilization rates from surfaces such as glass plates or leaves generally are higher than those from soil surfaces. For instance, the volatilization rate of methyl parathion (vp 2.0×10^{-5} mmHg) from a glass surface and from a moist soil column was 0.44 μg/cm^2/hr and 0.03 μg/cm^2/hr, respectively (Spencer et al. 1979).

2.1.3 Leaching

The term leaching refers to the movement of pesticide residues into the soil profile, and potentially to groundwater via percolating water. The exposure of biota to pesticides in groundwater may be of concern, depending on the toxicity and concentration of the chemicals. The factors that influence leaching potential include chemical properties, sorption behavior, and persistence, in addition to environmental variables such as rainfall and soil porosity (IAEA 1991).

Some OP compounds, such as chlorpyrifos, are relatively non-mobile in the soil profile, because they bind strongly to organic matter (K_{oc} = 8753). A study conducted by McCall et al. (1985) showed that chlorpyrifos was confined to the upper 5 cm of several soil types after its elution with 20 cm of water. Those observations were also confirmed by Oliver et al. (1987), wherein the residues of the pesticide were confined to the upper 12 in. of soil, in trials at several locations performed under field conditions. On the other hand, residues of fenamiphos were found at a concentration of 1,000 μg/L in groundwater from a Perth suburb (Western Australia), when analyzed after extensive plant death occurred in local gardens irrigated with groundwater (Appleyard 1995). A study on the mobility of pesticides, conducted by the Soil Conservation Service of the US Department of Agriculture, has ranked most pesticides tested as having a small to medium leaching potential (Table 1; Racke 1992).

2.1.4 Runoff

The transport of pesticides over the surface of treated fields via moving water or sediment is an important route of pesticide dissipation. Runoff is of concern, because aquatic non-target organisms could be affected by pesticides that move to waterways from sites of application. Generally, pesticides with a solubility greater than 10 mg/L will move mainly in the solution phase, and less soluble pesticides will move mainly by sorption to eroding soil particles (Wauchope 1978). The movement of fairly soluble compounds is driven chiefly by degree of solubility and environmental factors such as rainfall and hydraulic conductivity of the soil. The movement of more highly sorbed chemicals occurs mainly as an erosion-linked process and is strongly dependent on erosion management practices (Racke 1992). The quantity of chlorpyrifos transported by runoff to a pond, after three continuous applications, in a US (Iowa) study, reached only 0.03%. Similarly, approximately 0.1% of a granular diazinon applied was present in runoff from corn fields (Ritter et al. 1974). Data from turf (an irrigated environment) indicated that erosion did not occur because there was an absence of soil particles to which strongly sorbed OPs (e.g., chlorpyrifos) could adhere for transport (Watschke and Mumma 1989). However, other highly water-soluble OP compounds such as fenamiphos (400 mg/L) are able to migrate to surface water bodies because of low binding to the soil matrix (Cáceres et al. 2002).

2.2 Chemical Processes

There are several chemical reactions that influence the persistence of OP pesticides in the environment. Each will be reviewed, in turn, below.

2.2.1 Hydrolysis

Hydrolysis, which may result in detoxification, is an important process in mitigating the environmental persistence of organophosphate pesticides. The phosphate ester bond is a weak link in the molecule and is prone to cleavage. In some cases, hydrolysis can occur at several locations in an OP pesticide's structure; the most common reaction involves cleavage at the phosphate ester linkage as a result of base catalysis (Racke 1992). Both pH and temperature are important factors in the hydrolysis of OP compounds. OP structures usually undergo hydrolysis at pH 7 and 25°C; however, this process also occurs at a pH of 5–9 in groundwaters. In general, the higher the pH, the higher the rate of hydrolysis for many OP insecticides (Ragnarsdottir 2000).

Some examples of hydrolysis half-lives for OP compounds are shown in Table 1. From these data, it is evident that there is a wide variation in the pH-dependent hydrolysis rate of OP compounds. For instance, chlorpyrifos has hydrolysis half-lives of 72 and 16 d at pH values of 7 and 9, respectively. On the other hand, diazinon is acid hydrolyzed rapidly, with a half-life of 2 d at a pH of 5.0.

2.2.2 Oxidation/Reduction

OP insecticides may be oxidized or reduced to form metabolites. Oxidation and reduction reactions may proceed from the action of sunlight, from interaction with soil components or as an indirect consequence of microbial activity. For example, the oxidation of phosphorothionates such as parathion or chlorpyrifos results in the formation of oxon metabolites, which usually are less stable than the parent compounds (Racke 1992). However, in some phosphoramidates, such as isofenphos, the oxons are very persistent in soils (Felsot 1984). Some OPs, such as parathion, can be either reduced (to amino parathion) or oxidized (to paraoxon) (Katan and Lichtenstein 1977). Reductive dechlorinations can also occur under anaerobic conditions, as in the case of coumaphos and tetrachlorvinphos (Beynon and Wright 1969; Shelton and Karns 1988) and parathion (Sethunathan 1973; Sethunathan et al. 1977). The environmental significance of OP pesticide oxidation is that metabolites produced, such as oxons from phosphorothioates, may be more toxic than the parent compounds (Chambers and Levi 1992), but less stable than the parent compound.

2.2.3 Photolysis

Photolysis is an important process for the dissipation of OP compounds in surface waters and in the atmosphere. Photolysis occurs when chemicals are bombarded by solar light having wavelengths >290 nm. Photochemical transformation can occur by direct or indirect photolysis. Direct photolysis, a first-order process, occurs when the chemical absorbs light and then undergoes a transformation reaction from an excited state. Such transformations include rearrangement, dissociation, and oxidation. Indirect photolysis, a second-order rate process, occurs when substances naturally present in aquatic environment absorb sunlight to form excited chemical species or radicals, which then react with the pollutant chemical (Ragnarsdottir 2000). The transformation of disulfoton to disulfoton sulfoxide, parathion to paraoxon, and chlorpyrifos to 3,5,6-trichloro-2-pyridinol is catalyzed by photolytic reactions (Racke 1992). Another example is the isomerization of chlorfenvinphos z to chlorfenvinphos e, which is less toxic, a reaction mediated by the influence of sunlight (Cáceres 1995).

2.3 Biological Processes

Biological process includes enzyme-catalyzed transformation of OP compounds and the buildup of chemicals in the food chain. The degradation of pesticides in the subsoil, vadose zone, primarily results from microbiological action. OP compounds such as isofenphos, chlorpyrifos, parathion, coumaphos, and monocrotophos undergo degradation in this zone (Racke and Coats 1987; Singh and Walker 2006).

The rate at which biodegradation proceeds is a function of the microbial biomass present and the concentration of the OP, under given environmental conditions. Microorganisms are capable of using chemical substrates as an energy source. In such cases, biodegradation rates are a function of cell growth rate (Ragnarsdottir 2000). There are two classes of microbial pesticide degradation: cometabolism and catabolism. Cometabolism often results in incomplete microbial degradation of the pesticide and subsequent transient accumulation of some of its metabolites (Alexander 1981). Microorganisms that initiate the cometabolic degradation process obtain no benefit from their activities, and thus, this process is often termed incidental degradation. For example, parathion in mixed cultures of soil microorganisms was quantitatively transformed to aminoparathion by cometabolism within 12 hr (Katan and Lichtenstein 1977). Likewise, in a flooded soil, parathion is cometabolically hydrolyzed by microorganisms, and p-nitrophenol formed by hydrolysis, which is then utilized by microorganisms as an energy source for their proliferation. In flooded soils, parathion is transformed into aminoparathion by nitro group reduction or to p-nitrophenol by hydrolysis, or both (Barik and Sethunathan 1978; Sethunathan et al. 1977). In natural systems, these metabolites do not accumulate, but are degraded by other microorganisms, which do not obtain any corresponding benefit.

In catabolic processes, a parent pesticide or a primary metabolite is degraded with a concomitant benefit to the microbe, because it utilizes the compound as a carbon/energy source or nutrient. This occurs with diazinon and parathion, which were hydrolyzed by *Flavobacterium* sp. These bacteria used the hydrolysis products from diazinon and parathion as sole carbon sources (Sethunathan and Yoshida 1973). One of the enzymes evolved in this process (parathion hydrolase) has been isolated from *Pseudomonas diminuta* (Mulbry et al. 1986). The ability of microorganisms to metabolize pesticides may also result in an adaptive event. It has been noted that repeated exposure of soil or aquatic microorganisms to some OP pesticides results in the proliferation of pesticide-degrading microorganisms, mainly bacteria, that have enhanced biodegradation capacity. Some OP compounds such as diazinon, coumaphos, fensulfothion, and chlorfenvinphos are known to be susceptible to this phenomenon (Sethunathan and Pathak 1972; Racke and Coats 1990).

Recently, results from several studies (Karpouzas and Walker 2000; Karpouzas et al. 2000, 2004a) demonstrated the capacity of some microorganisms (e.g., *Flavobacterium*, *Pseudomonas*, and *Sphingomonas*) to break down OP pesticides such as parathion, cadusafos, and ethoprophos. Pakala et al. (2007) isolated a bacterium capable of utilizing methyl parathion as a sole carbon and energy source. The bacterial strain belonged to the genus *Serratia*, based on a phylogram constructed using the complete sequence of its 16S rRNA. The major degradation products of methyl parathion were *p*-nitrophenol and methylthiophosphoric acid. These authors showed that the enzymes involved in the degradation were parathion hydrolase and *p*-nitrophenol hydrolase component "A."

Xu and coworkers (2007) reported that a bacterial strain of *Serratia* sp. transformed chlorpyrifos to 3,5,6-trichloro-2-pyridinol (TCP) and that a fungal strain *Trichosporon* sp. was capable of mineralizing TCP. These microorganisms were isolated from an activated sludge. It was observed that the cultures completely mineralized 50 mg/L chlorpyrifos within 18 hr at 30°C and pH 8, using a total inocula of 0.15 g/L biomass. Jiang et al. (2007) demonstrated the simultaneous biodegradation of methyl parathion and carbofuran by a genetically engineered microorganism (GEM). This GEM was constructed by random insertion of a methyl parathion hydrolase gene (*mpd*) into the chromosome of a carbofuran-degrading *Sphingomonas* sp. CDS-1 that had the mini transposon system. The constructed GEM was relatively stable, and cell viability and original degrading characteristics were unaffected compared to the original recipient CDS-1. The cells of this GEM could degrade methyl parathion and carbofuran efficiently at a relatively broad range of temperatures (20–30°C), initial pH values (6–9), and with an initial inoculation cell density of 10^5–10^7 CFU (colony forming units)/mL, even if an alternative glucose source existed.

Yang and collaborators (2007) isolated a soil bacterium, which was identified as *Bacillus* sp. using a 16S rRNA gene sequence and BIOLOG test. The bacterium was able to transform parathion and methyl parathion into amino derivatives by reducing the nitro group; this reaction was catalyzed by a nitroreductase. Zhang et al. (2006) reported the identification of a gene capable of encoding the organophosphorus hydrolytic enzyme responsible for degrading fenitrothion by a Gram-negative bacterium strain, *Burkholderia* sp. FDS-1. This bacterium was able to use fenitrothion as the sole carbon source for its growth. The gene responsible for encoding the enzyme was cloned and sequenced. The sequence was similar to *mpd*, a gene previously shown to encode a parathion-methyl-hydrolyzing enzyme in *Plesiomonas* sp. M6.

DebMandal and collaborators (2008) isolated two bacterial strains, *Bacillus licheniformis* and *Pseudomonas aeruginosa*, from water and fish intestine, respectively. These bacteria were able to degrade dimethoate. In the case of *P. aeruginosa*, four metabolites were formed, whereas total disappearance of the dimethoate occurred in the culture with *B. licheniformis*. Chu and collaborators (2006) reported the expression of an organophosphorus hydrolase OPHC2 in *Pichia pastoris*. The expression of the hydrolase was similar to the one observed in the original bacterium *Pseudomonas pseudoalcaligenes* C2-1. The purified enzyme exhibited broad temperature and pH stability. Its efficiency was tested with methyl parathion as the substrate.

A new fungal enzyme capable of hydrolyzing methyl parathion, parathion, paraoxon, coumaphos, demeton-S, phosmet, and malathion was obtained from *Penicillium lilacinum* BP303; this enzyme was purified and characterized by Liu and coworkers (2004). The purified enzyme was considered to be a novel organophosphorus hydrolase because of its accessibility to detergents, resistance to many metal ions, and broad pH and temperature optima. These findings are important, because they suggest the ability of some bacteria to degrade organophosphate insecticides potentially useful in bioremediation programs.

3 Soil Bioavailability of OP Compounds

Bioavailability represents the biological accessibility of a chemical for assimilation by organisms that could produce toxicity (Alexander 2000). Bioavailability is linked to several factors, including the nature of the organism, physicochemical properties of the contaminant, and type of environmental factors present (Juhasz et al. 2000). Bioavailability may be affected by soil characteristics such as pH and quantity and quality of organic matter and clay minerals present (Van der Wal et al. 2004). When a pesticide is applied to soil, it may be rapidly bound to mineral and organic matter through physical, chemical, and biological processes. The ability of the soil to release bound contaminants affects the degree to which these contaminants are subject to microbial degradation. The degradation rate, in turn, influences the success of contaminated soil cleanup when using contaminant-degrading microorganisms, a process known as bioremediation (Juhasz et al. 2000). Therefore, data on pesticide sorption and desorption rates are critical for predicting xenobiotic attenuation and bioavailability phenomena in ecosystems (Gao et al. 1998; Lawrence et al. 2000).

There are several methods used to measure the bioavailability of organic compounds in soil. The direct method is to expose a target organism to a xenobiotic and measure the resultant uptake of the compound, which usually results in accumulation and/or manifestation of an organism effect over a fixed period per unit mass of soil (Harmsen et al. 2005). Despite research being focused, in recent years, on the development of methods to measure bioavailability of organic compounds in soil (Reid et al. 2000; Semple et al. 2004; Lanno et al. 2004), there are as yet no standardized procedures available. Some microbiological methods have been used to measure biodegradability and as such bioavailability. Mineralization assays are often designed to measure ^{14}C parent compound conversion to $^{14}CO_2$ which provides a measure of the microbially available fraction of the contaminant (Hatzinger and Alexander 1995).

The availability of pesticides to microorganisms is important because it determines their potential for biodegradation, which influences the persistence of these xenobiotics in the soil environment (Table 2) and also the potential for their transformation and availability to plants and higher organisms. The bioavailability of organic chemicals to microorganisms is governed by the type of soil in which they reside, structural characteristics of the organic chemical, and properties of the microbes involved (Guerin and Boyd 1992; Feng et al. 2000). Karpouzas et al. (2000) demonstrated that *Pseudomonas putida* epI was able to degrade the OP compound ethoprophos in aged and fresh fumigated soils. Singh et al. (2003a), using a combination of X-ray diffraction (XRD) and infra-red (IR) analysis, have shown that fenamiphos, intercalated on cetyltrimethylammonium-modified montmorillonite (CTMA-Mt), was bioavailable to *Brevibacterium* sp. for degradation at sorption sites, without being desorbed into the solution phase.

Earthworms are in close contact with soil and the soil solution, and they consume large volumes of soil. Since bioavailability can be measured in terms of bioaccumulation and toxicological bioavailability (Lanno et al. 2004), earthworms are frequently used to assess the bioavailability of a wide range of organic soil pollutants, including pesticides (Meharg 1996; Morrison et al. 2000; Kelsey et al.

Table 2 Half-lives of selected OP compounds in soil

Pesticide	Half-life in soil (d)
Azinphos-methyl	10
Chlorpyrifos	30
Coumaphos	300
Diazinon	40
Dichlorvos	0.5
Dicrotophos	20
Dimethoate	7
Disulfoton	30
Ethoprop	25
Fenamiphos	50
Fenthion	34
Fonofos	40
Isofenphos	150
Malathion	1
Methamidophos	6
Monocrotophos	30
Parathion ethyl	14
Profenofos	8
Temephos	30

Adapted from Kearney et al. (1986) and National Pesticide Information Centre (1994)

2005). Detrimental effects in worms caused by exposure to OP pesticides such as diazinon and chlorpyrifos have been reported by several researchers (Stenersen et al. 1992; Booth and O'Halloran 2001; Rao et al. 2003; Mosleh et al. 2003). Avoidance behavior of earthworms has been reported in several studies involving the toxicity testing of pesticides such as glyphosate, imidacloprid, and benomyl (Christensen and Mather 2003; Capowiez et al. 2004; Verrell and Buskirk 2004). The toxicity of OP pesticides to worms appears to be quite variable. For example, the LC_{50} of chlorpyrifos for *Eisenia foetida*, as determined by Rao et al. (2003), was just 2.33 mg/kg, whereas the LC_{50} of profenofos on *Aporrectodea caliginosa* corresponded to 127 mg/kg (Mosleh et al. 2003), and the LC_{50} established for diazinon on *A. caliginosa* was 102 mg/kg (Booth and O'Halloran 2001). Mosleh and coworkers (2003) found that some biochemical responses in the earthworm *A. caliginosa* are mediated by pesticides such as aldicarb, cypermethrin, profenofos, chlorfluazuron, atrazine, and metalaxyl. Those responses included an increase in the levels of transaminases and phosphatases.

Historically, soxhlet extraction has been regarded as an exhaustive way to remove and measure the amount of a contaminant in soil. However, recent information demonstrates that exhaustive extraction gives little information on actual compound bioavailability and can result in inappropriate estimation of exposure and risk to susceptible populations (Kelsey and Alexander 1997). Several mild solvent extractions of organic pollutants in soil were performed, and amounts extracted were compared with uptake by particular organisms. Most of this research was focused on persistent organic pollutants such as organochlorines, polyaromatic hydrocarbons (PAHs), and polychlorinated biphenyls (PCBs). The extractants for these substances included methanol–water, ethanol, *n*-butanol, and acetone–water (Krauss et al. 2000; Liste and Alexander 2002; Thiele-Bruhn and Brummer 2004). Liste and Alexander (2002) showed a good correlation between the amounts of phenanthrene and pyrene biodegraded in soils, with or without plants, and the amounts extractable by *n*-butanol. Similar extraction studies to glean information on relative bioavailability were performed on several herbicides (atrazine, simazine, and alachlor), as well (Gan et al. 1999; Barriuso et al. 2004; Reginato et al. 2006). Barriuso et al. (2004) and Reginato et al. (2006) have demonstrated that the amount of atrazine and simazine extracted from aged soils by 0.01 M $CaCl_2$ and aqueous methanol (80:20 v/v methanol/water) were highly correlated to the amounts of simazine mineralized by a simazine degrading bacterium, *Pseudomonas* sp. strain ADP. These authors stated that this extraction technique may be useful for determining the bioavailability of other s-triazine compounds in soils. However, information on the bioavailability of OP compounds in soil is scarce and only few studies, with diazinon, for example (Boxall et al. 2006), have been reported. Results suggest that it is feasible to predict the bioavailability of these chemicals in soil by solvent extraction, but the conclusions drawn are not consistent. This is partly because extractants are compound selective and target organisms specific in their response, which means that results from one study with a specific compound and organism, cannot be easily extrapolated to another compound or organism.

4 OP Pesticides and Health

The inhibition of AChE at critical neuronal sites in mammals and insects by OP insecticides can result in severe poisoning or death. The cause of death from OP poisoning in mammals is usually respiratory failure (Casida and Quistad 2004). These pesticides inhibit not only acetylcholinesterase but also the function of carboxyl ester hydrolases, such as chymotrypsin, plasma, or butyrylcholinesterase, plasma and hepatic carboxylesterases (aliesterases), paraoxonases, and other nonspecific esterases within the body (Abdollahi et al. 2004). Several studies have reported oxidative stress induced by OPs in humans (Banerjee et al. 1999; Ranjbar et al. 2002; Dantoine et al. 2003). Lipid peroxidation was found in human erythrocytes after exposure to OP pesticides (Gultekin 2000; Dantoine et al. 2003). OP-induced seizures have been reported to be associated with oxidative stress (Gupta 2001). Pazy Miño et al. (2002) performed a cytogenetic monitoring in a population occupationally exposed to pesticides, including OP insecticides such as fenamiphos and profenofos. This study was performed at a flower plantation located in Quito, Ecuador. It was observed that 88% of the exposed individuals showed low levels of erythrocyte acetylcholinesterase, 28 Units/mL being the optimal level in the blood. A highly significant correlation was observed between levels of erythrocyte acetylcholinesterase and the percentage of structural chromosome alterations, indicating that individuals with low erythrocyte acetylcholinesterase levels also show an increase in chromosomal aberrations. The authors concluded that, using both analyses, it was possible to estimate the damage produced by exposure to OP pesticides.

Acute poisoning with organophosphate-based pesticides is a notable cause of morbidity and mortality in the developing world (Blain 2001). The potential for frequent low-level exposure to OP pesticides exists for many occupational groups including agricultural workers, sheep dippers, and pesticide sprayers. Some study results (Buchanan et al. 2001; Jamal et al. 2002; Farahat et al. 2003) suggested that long-term effects on the central and peripheral nervous systems may be associated with frequent but low-level exposure to OP insecticides. These neurological effects were different from those associated with the delayed neuropathy known to result from acute poisoning by certain OP compounds. These additional neurotoxic effects in humans ranged from neurobehavioral and electroencephalographic changes to increases in the variability of action potential latencies in skeletal muscles. Farahat et al. (2003) conducted a monitoring study in Egypt on farmers working at cotton farms for a 3-mon period. Most of the workers said that they never used protective clothes and all had applied pesticides for at least 3 yr. The OP pesticides cited as having been used included curacron, dursban, hostathion, thimet, profenofos, chlorpyrifos, triazophos, and phorate. A control group was also monitored. Exposed workers exhibited lower performance than control groups on most of the neurobehavioral tests, which included verbal abstraction, visuomotor speed, problem solving, attention, and memory. Symptoms of dizziness and numbness were significantly higher in the exposed workers than in the control group. Other effects observed in exposed workers were blurred vision, paresthesia, headache, vertigo, superficial sensory loss, trophic and vasomotor changes, and decreased ankle and deep reflexes. Jamal and collaborators (2002) conducted a study to determine the neurological and neurophysiological effects on sheep farmers and dippers exposed to OP pesticides in the UK, based on nerve conduction measurements or presence of neurological signs. Clinical neuropathy increased in 52% of the exposed group. Sensory abnormalities were found more often than motor deficits. Small diameter nerve fibers were affected more than large fibers. The authors concluded that the use of OP compounds should be urgently reviewed and need for health and safety requirements should be evaluated.

5 Fenamiphos

Fenamiphos (ethyl 4-methylthio-*m*-tolyl isopropylphosphoramidate) (Fig. 2), is an OP nematicide used to control a wide variety of nematode (roundworm) pests. This pesticide is used on a variety of plants including tobacco, turf, bananas, pineapples, citrus, and other fruit vines, some vegetables, and grains. Fenamiphos, as is typical of other OP compounds, blocks the enzyme acetylcholinesterase in the target pest. This pesticide also exhibits secondary activity against other invertebrates such as sucking insects and spider mites. It is available in emulsifiable concentrate, granular, or emulsion formulations (Tomlin 2000). The physicochemical properties of this pesticide are summarized in Table 3.

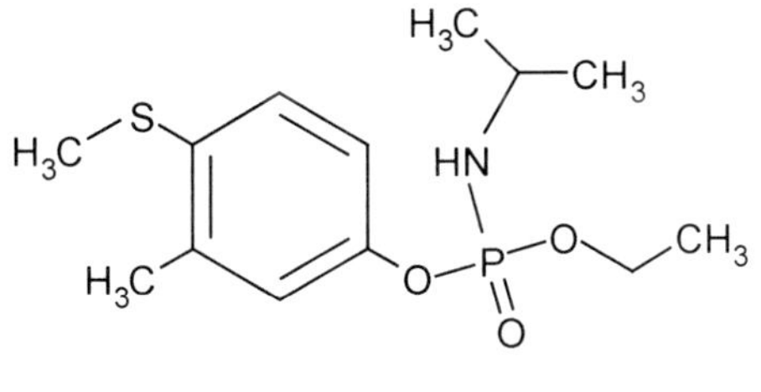

Fig. 2 Fenamiphos

The pesticide is registered for use in more than 60 countries. It provides effective control of free living root-knot and cyst-forming nematodes. The main target species are *Meloidogyne* spp., *Pratylenchus* spp., *Radopholus similes*, *Rotylenchus* spp., *Rotylenchus reniformis*, *Heterodera* spp., and *Xiphinema* spp. Fenamiphos provides crop protection against nematode damage through systemic activity in the plant and also via contact action in the soil. This pesticide was commercialized by the company Bayer under the name of Nemacur and is formulated as a granular product (5, 10, 12, and 15 GR) or an emulsifiable concentrate at 400 g active ingredient per liter. This pesticide may be applied pre-plant in established crops or in seedbeds and nurseries. For effective control, it is important that Nemacur formulations are incorporated into the soil in root-growth zones, because it is the roots that are exposed to nematodes. Fenamiphos is predominantly used in tropical areas in the USA, South America, South Africa, Spain, Australia, Costa Rica, and Italy.

Table 3 Physicochemical properties of fenamiphos

Physical/chemical property	Value	Temperature (°C)	Source
Water solubility	0.4 g/L	20	Tomlim (2000)
Solubility in hexane	10–20 g/L	20	Patrick et al (2001)
Molecular weight	303.4		Tomlim (2000)
Melting point	49°C		Tomlim (2000)
pKa dissociation constant	10.5		National Library of Medicine (2004)
Log P (octanol–water)	3.23	20	National Library of Medicine (2004)
Specific gravity/density	1.191	23	Tomlin (2000)
Vapor pressure	0.12 mPa at 20°C	25	Extoxnet (1996)
Henry's Law constant	9.1×10^{-5} Pa m^3/mol	20	Tomlin (2000)
Atmospheric OH rate constant	7.77E-11 cm^3/molecule/sec	25	National Library of Medicine (2004)

5.1 Environmental Fate

5.1.1 Breakdown in Water and Soils

Data that address surface water monitoring for fenamiphos and its metabolites is scarce. In 2000, the EPA and the US Geological Survey conducted a study to monitor pesticide residues in 12 drinking water reservoirs (Patrick et al. 2001). Residues of fenamiphos and its degradation products, fenamiphos sulfoxide (FSO) and fenamiphos sulfone (FSO$_2$), were found in three reservoirs at concentrations varying from 0.005 to 0.033 µg/L. Fenamiphos applied to fields may reach surface water bodies via runoff from the treatment site and via spray drift. The results of this study support the conclusion that fenamiphos and its degradation products may enter surface water and be subsequently found in drinking water, in areas where the community water system is in close proximity to use areas (Patrick et al. 2001).

Under laboratory conditions, fenamiphos was found to be stable in neutral water held in the dark, but the pesticide disappeared rapidly under acidic and alkaline conditions. The half-life of fenamiphos in a neutral solution was 4 hr and it can be photolytically degraded to FSO in approximately 3.23 h (Patrick et al. 2001). The pesticide was susceptible to degradation when exposed to artificial light (Extoxnet 1996). This pesticide undergoes degradation on soil surfaces and has an average bioactivity of 4 mon (Tomlin 2000).

According to sorption and leaching data, fenamiphos can be classified as a low soil mobility chemical (Tomlin 2000). However, because of its chemical characteristics, fenamiphos, and its major degradation products (FSO and fenamiphos sulfone) have the potential to leaching to groundwater in vulnerable areas. Residues of these chemicals were found in groundwater samples from California, Georgia, and Florida in the USA (Patrick et al. 2001), as shown in Table 4.

Table 4 Residues of fenamiphos and its degradation products found in US groundwater

Place	Well type	Crop treated	Concentration rate (μg/L)
California	Monitoring	Grapes	0.05 (F) 0.06–2.13 (FSO) 0.53 (FSO$_2$)
Georgia	Monitoring	Tobacco	0.0 (F) 0.04–0.05 (FSO) 0.0 (FSO$_2$)
Florida	Monitoring	Citrus	0.10–0.58 (F) 0.13–83 (FSO) 0.14–3.3 (FSO$_2$)
Florida	Monitoring/irrigation	Golf course	0.03–0.71 (F) 0.2–0.75 (FSO) 0.1 (FSO$_2$)

F = fenamiphos; FSO = fenamiphos sulfoxide; FSO$_2$ = fenamiphos sulfone
Adapted from Patrick et al. (2001)

The sorption of fenamiphos and its metabolites has been studied in detail in soils from Australia (Oliver et al. 2003) and Ecuador (Cáceres et al. 2002; Cáceres et al. 2008a). However, the information on sorption of fenamiphos and its metabolites is scarce in soils from other countries. A summary of the K_d values determined in surface soils from these countries is presented in Table 5. Simon et al. (1992) reported the sorption coefficient of fenamiphos and its metabolites for different soils around the world (Table 6).

Ou and Rao (1986) reported the degradation and metabolism of fenamiphos in soils from the USA. The pesticide was oxidized to FSO and then slowly degraded to fenamiphos sulfone. In addition, FSO was also hydrolyzed to fenamiphos phenol. Fenamiphos disappeared faster in moist soil than in dry soil. The half-life for total toxic residues of the compound ranged from 38 to 67 d. Kookana and coworkers (1997) studied the transformation and degradation of fenamiphos and its metabolites in sandy soils from Australia. In this study, the pesticide exhibited a half-life of 50 d in surface soils and 140 d in subsurface soils. The parent compound was rapidly oxidized to FSO and then at very low rates to fenamiphos sulfone. Simon et al. (1992) examined the influence of soil properties on the degradation of fenamiphos in 16 soils from around the world. They found that the half-life in soils from temperate regions, incubated at 16°C, varied from 29.2 to 166.7 d, whereas the half-life of the compound in soils from tropical regions incubated at 28°C varied from 14.1 to 52.6 d.

Cáceres et al. (2008b) studied the degradation of fenamiphos in different soils from Australia, India, and Ecuador. The soils were incubated at different temperatures to simulate different climatic conditions. The half-life of fenamiphos varied from 2.42 to 7.18 d at 18°C; at 25°C the half-life ranged from 2.45 to 3.18 d, and at 37°C the half-life varied from 1.16 to 2.49 d. FSO was identified as the major

Table 5 Sorption coefficients (K_d) for fenamiphos, F sulfoxide, and F sulfone for soils from Australia and Ecuador

Soil classification	Fenamiphos K_d	F sulfoxide K_d	F sulfone K_d	Source
Typic Durothod	38.68	nd	nd	Oliver et al. (2003)
Xeric Quartzipsamment	4.60	0.07	0.48	Oliver et al. (2003)
Typic Durothod	25.86	3.01	2.82	Oliver et al. (2003)
Typic Psammaquent	21.90	1.65	4.04	Oliver et al. (2003)
Petrocalcic Xerochrept	8.86	1.25	0.53	Oliver et al. (2003)
Xeric Quartzipsamment	7.37	0.81	2.09	Oliver et al. (2003)
Hapludolls[a]–Fluventic Europepts	10.77	1.79	5.28	Cáceres et al. (2002)
Ustipsamments[a]	12.75	2.47	6.20	Cáceres et al. (2002)
Vertic Ustropepts[a]	5.66	1.55	3.67	Cáceres et al. (2002)
Andic Eutropept[a]	8.94	0.77	2.81	Cáceres et al. (2002)
Kanhapludalf[a]	7.91	1.06	2.83	Cáceres et al. (2002)
Tropofluvents[a]	14.31	4.00	8.79	Cáceres et al. (2002)
Dystrandepts[a]	14.19	0.95	1.62	Cáceres et al. (2008b)
Durustols[a]	4.07	0	0	Cáceres et al. (2008b)
Dystrandepts[a]	12.57	1	1.71	Cáceres et al. (2008b)
Haplustolls[a]	8.95	2.27	2.66	Cáceres et al. (2008b)

The data values are single point measurements
nd = not determined
[a]Ecuadorian soils

Table 6 Adsorption constants (K_{oc}) for fenamiphos and its major degradation products in soils around the world

Location	% clay	% silt	% sand	Fenamiphos (K_{oc})	Fenamiphos sulfoxide phenol (K_{oc})	Fenamiphos sulfone phenol (K_{oc})
Canada	22.7	47.6	29.7	297.9	82.6	133.1
Sweden	8.5	9	82.5	279.7	12.5	49.4
Puch, Germany	14.1	74.9	11	111.6	2.8	47.2
Speyer, Germany	4.5	14.1	81.4	178.4	44.5	57.7
Netherlands	19.3	58	22.7	380	41.3	66.1
France	28.1	39.9	32	159.5	257	40.4
Indiana, US	12	25.6	62.4	234.7	6.4	103.4
Nebraska, US	27	69.8	3.2	312.6	118.3	152.9
Japan	9.8	48.1	42.1	76.2	27.2	31
Florida, US	1.3	3.3	95.4	226	132.1	146.8
Costa Rica	29.6	41.6	28.8	363.4	165.6	206.6
Brazil	44.4	24.4	31.2	140.5	56.3	68
Thailand	55.3	43	1.7	1432	102.4	198.5

Adapted from Simon et al. (1992)

degradation product. However, traces of the other metabolites (FSO_2 and the corresponding phenols) were detected. The degradation was faster in alkaline soils than in acidic and neutral soils.

The mean depth to which fenamiphos and its metabolites leached, when studied under field conditions on sandy soils in Australia, was 28 cm (Kookana et al. 1995). Leaching studies conducted in columns under controlled conditions, using ^{14}C fenamiphos, demonstrated that the pesticide was relatively mobile, with 16.2–63.8% of the applied radioactivity found in the leachate. The main metabolites found in the leachate were FSO and FSO_2, which were more mobile than the parent compound (Patrick et al. 2001). It has been reported that inappropriate disposal of fenamiphos residues caused groundwater contamination and that the pesticide half-life in an aquifer under anaerobic conditions may reach 1,000 yr (Franzmann et al. 2000).

The degradation of fenamiphos in soil is also influenced by microbial activity. A fraction of the soil biota present in the soil can develop an enhanced ability for degrading pesticides, after successive applications to the soil. This microorganism-mediated process has been described as enhanced or accelerated biodegradation (Sethunathan and Pathak 1972; Singh et al. 2005). The efficacy of fenamiphos can be extensively reduced by enhanced biodegradation. Enhanced degradation can occur not only from repeated applications but may also occur after a single pesticide application. Ou (1991) reported enhanced degradation of fenamiphos in soil collected from a site planted with potatoes and treated with one application of fenamiphos, in the USA. The half-lives for total toxic residues of fenamiphos in these soils, collected 2 and 3 yr after field application of the nematicide, were 22 and 89 d, respectively, whereas the half-life values for the corresponding untreated control soils were 131 and 130 d. Ou et al. (1994) studied the degradation of fenamiphos in a Florida soil that had a history of 20 yr of continuous applications for controlling nematodes in turfgrass golf courses. It was observed that the efficacy of the pesticide was reduced when it was applied once or twice a year for extended periods. The half-life values for total toxic residues were very short, with values ranging from 0.9 to 4.2 d. The main metabolite observed was FSO, but FSO_2 was not detected. In this study it was demonstrated that the main pathway of degradation of fenamiphos was via oxidation to fenamiphos sulfoxide, which was then quickly degraded to CO_2 and water. Chung and Ou (1996) also demonstrated the enhanced biodegradation of FSO and FSO_2 in soils with a long history of fenamiphos application. Anderson et al. (1998) reported a more rapid degradation of fenamiphos in treated vs. control soils from Costa Rica banana plantations that had not previously been exposed to the nematicide.

Information on the microbial degradation of fenamiphos in soils is scant. A microbial consortium from the USA capable of degrading fenamiphos in liquid medium was isolated by Ou et al. (1994). Another microbial consortium able to degrade fenamiphos was isolated from Australian soils. Molecular analysis of the bacterial populations showed a similarity to some of the genera *Pseudomonas*, *Flavobacterium*, and *Caulobacter*, and it was observed that the degradation of the pesticide was faster in soils with high pH (Singh et al. 2003b). Karpouzas et al. (2004b) also demonstrated enhanced biodegradation of fenamiphos in soils collected from potato fields having a 10-yr history of treatment with the nematicide. The degradation of fenamiphos in these soils was faster than that in control soils or in acclimatized soils fumigated with the antibiotic chloramphenicol, which inhibited the microorganisms responsible for enhanced fenamiphos degradation. Megharaj et al. (2003a) demonstrated the involvement of microbes in the degradation of fenamiphos in a local South Australian soil spiked with fenamiphos and incubated under laboratory conditions. About 80% of initially spiked fenamiphos (25 mg/kg soil) was degraded in non-autoclaved soil, compared to only 9% in autoclaved soil, during the 20-d incubation period, which clearly demonstrates the microbial role in fenamiphos degradation. The first report of isolation of a pure bacterium with an exceptional ability to degrade fenamiphos is from Megharaj et al. (2003a). This bacterium, isolated from an Australian soil and identified as *Brevibacterium* sp. MM1, was able to hydrolyze not only fenamiphos but also its oxides in soil and groundwater. Recently, Cáceres et al. (2009) reported the isolation of another pure bacterium able to hydrolyze fenamiphos and its oxides. This bacterium was isolated from turf soil and was identified as *Microbacterium esteraromaticum*, based on its 16S rRNA gene sequence.

Cáceres et al. (2008d) evaluated the ability of different species of five green algae (*Scenedesmus* sp. MM1, *Scenedesmus* sp. MM2, *Stichococcus* sp., *Chlorella* sp., and *Chlamydomonas* sp.) and five cyanobacteria (*Nostoc* spp. MM1, MM2, and MM3, *Nostoc muscorum*, and *Anabaena* sp.) to degrade fenamiphos. All the species tested were able to degrade fenamiphos to its primary oxidation product FSO, whereas the majority of these cultures were able to hydrolyze FSO to fenamiphos sulfoxide phenol (FSOP). Fenamiphos sulfone phenol (FSO_2P), FSOP, and FSO were detected in the culture extracts of these algae and cyanobacteria. If one considers the hydrolysis of fenamiphos as a detoxification step, the ability of these algae and cyanobacteria to hydrolyze fenamiphos and its toxic oxidation products, can gainfully be used to detoxify fenamiphos contaminated soils and water. This report is the first that treats the biodegradation of a toxic pesticide, fenamiphos, by the cyanobacteria.

Another important phenomenon observed in soil microorganisms exposed to pesticides is the development of cross-enhancement or accelerated biodegradation of related compounds. Several studies have demonstrated the cross-adaptation of microorganisms and enhanced degradation of related OP compounds (Sethunathan et al. 1977; Karpouzas et al. 2005). Singh and coworkers (2005), tested the potential of chlorpyrifos-degrading microorganisms to degrade related OP compounds, including fenamiphos. But, no cross-degradation of fenamiphos by chlorpyrifos-degrading bacteria was observed. Megharaj et al. (2003b) found that a fenamiphos-degrading bacterium, *Brevibacterium* sp., could not degrade other OP compounds such as methyl parathion. However, it will be useful to further investigate the capability of this microorganism to degrade other, similar OP compounds.

5.1.2 Metabolism in Plants

The metabolism of fenamiphos in beans, tomato, peanuts, and potato was studied by Waggoner in 1972. The seeds of beans (*Phaseolus vulgaris*), rutgers tomato (*Lycopersicon esculentum*), and peanuts (*Arachis hypogaea*) were germinated and planted in sand. After 4 wk, the beans were treated with ^{14}C fenamiphos, whereas the tomato and peanuts plants were exposed to the radiolabeled pesticide after 6 wk.

In addition, potato (*Solanum tuberosum*) was grown and treated with the pesticide at 6 wk. The plants were injected in the main stem with 20–50 μL of an ethanol:water radioactive solution. The treated plants were harvested at 0, 7, 14, and 28 d. It was observed that the pesticide was broken down within the plants. Two metabolites were detected: ethyl 4-(ethylsulfinyl)-*m*-tolyl isopropylphosphoramidate and ethyl 4-(methylsulfonyl)-*m*-tolyl isopropylphospharamidate. These metabolites were also active inhibitors of the acetylcholinesterase.

The World Heath Organization (WHO), in its 1974 publication, cited results provided by Khasawinah (in reports from the manufacturer Chemagro). These results were from studies of fenamiphos metabolism in carrots, cabbages, soybeans, and tobacco plants that were conducted using both soil treatment and stem injection. The research was performed using ^{14}C-labeled fenamiphos at three different positions (^{14}C ring, ^{14}C 1-ethyl, and ^{3}H methylthio). Carrots absorbed and metabolized fenamiphos and ^{14}C was detected in its foliage. The ^{14}C was probably incorporated into lipids, pigments, and sugars. Water-soluble residues like conjugates of the sulfoxide and sulfone phenols were also detected. The parent compound was not detected. The residues of sulfoxide and sulfone were below 0.06 mg/kg in carrots and foliage. The results observed in tomatoes were similar. No parent compound was detected at 40 d. Sulfoxide and sulfone residues reached 0.2 and 0.3 mg/kg after 40 d of treatment. In the experiment with head cabbage, a polar metabolite with a phosphoramidate structure was detected. In tobacco, the formation of the sulfone reached the highest concentration in less than 1 wk after treatment. In soybeans, besides the major metabolites, sulfoxide and sulfone, a glucose conjugate of sulfone phenol was detected.

5.2 Ecotoxicology

The toxicity of fenamiphos has been studied in aquatic and terrestrial organisms. Recently, Cáceres et al. (2008c) determined the toxicity of fenamiphos and its metabolites to the aquatic alga *Pseudokirchneriella subcapitata* and the terrestrial alga *Chlorococcum* sp. The toxicity followed the order fenamiphos phenol > fenamiphos sulfone phenol > fenamiphos sulfoxide phenol > fenamiphos. The most common fresh water species used in acute toxicity tests is *Daphnia magna*. These tests are conducted under laboratory conditions and are designed to estimate the quantity of a substance in water (mg/L) required to immobilize or kill 50% of an exposed population (LC_{50}). The LC_{50} of fenamiphos to *D. magna* correspond to 0.0019 mg/L (Tomlin 2000). The information on the toxicity of fenamiphos metabolites to *D. magna* is scarce and, therefore, studies to fill this gap are required (Patrick et al. 2001). The toxicity of the pesticide could vary among species, and therefore it is also important to conduct toxicity tests on different water flea species. Recently, Cáceres et al. (2007) determined the toxicity of fenamiphos and its metabolites to *Daphnia carinata*. The toxicity followed the order fenamiphos > fenamiphos sulfone > fenamiphos sulfoxide.

A fresh water aquatic invertebrate life-cycle test is required by many pesticide regulators. In this test *D. magna* are exposed for 21 d to the target compound, such as fenamiphos, under laboratory conditions. The test is designed to estimate the highest quantity of the xenobiotic in water (mg/L) that does not affect the reproductive capability of the invertebrates (no observable effect concentration; NOEC), and the lowest quantity that does affect the reproductive capability (lowest observable effect concentration; LOEC). The NOEC for fenamiphos in *D. magna* was 0.12 ppb and the LOEC was 0.24 ppb (Patrick et al. 2001). Acute toxicity tests for fenamiphos in marine and estuarine organisms have commonly been conducted using sheepshead minnow, mysid shrimp, and eastern oyster. There are a few data available on the toxicity of fenamiphos to these organisms: the LC_{50} of fenamiphos is 1.65 ppb to eastern oyster (*Crassostrea virginica*), 17 ppb for Sheepshead minnow (*Cyprinodon variegatus*), and 6.2 ppb to mysid shrimp (*Mysidopsis bahia*) (Patrick et al. 2001).

Toxicity studies with freshwater fish (96 hr) usually are conducted with Rainbow trout and Bluegill fish. Patrick et al. (2001) reviewed the acute toxicity to fresh water fish using fenamiphos technical, fenamiphos sulfoxide, fenamiphos sulfone, and the commercial formulation, Nemacur 3 and 10G, as summarized in Table 7. In early life stage tests, using Rainbow trout, the NOEC and LOEC values for fenamiphos technical were 0.0038 and 0.0074 mg/L, respectively (Patrick et al. 2001). Among the terrestrial invertebrates, *E. foetida* has been recognized as a representative test organism to assess the toxicity of xenobiotics to earthworms and other soil invertebrates, primarily because the growth and reproduction of this species is well documented (Vermeulen et al. 2001; Landrum et al. 2006). The LC_{50} of fenamiphos for this organism corresponded to 795 mg/kg in soil (Tomlin 2000). However, information regarding the harmful effect on these invertebrates, after long-term soil exposure to fenamiphos, is poorly understood, and therefore, research on

Table 7 Fresh water fish acute toxicity for fenamiphos and its metabolites

Species	% active ingredient	Formulation/ chemical	LC_{50} (ppb)	Toxicity category
Blue gill sunfish (*Lepomis macrochirus*)	88	Technical	9.5	Very highly toxic
Blue gill sunfish (*L. macrochirus*)	81	Technical	17.7	Very highly toxic
Rainbow trout (*Oncorhynchus* sp.)	81	Technical	72.1	Very highly toxic
Blue gill sunfish (*L. macrochirus*)	36	Nemacur 3	4.5	Very highly toxic
Rainbow trout (*Salmo gairdneri*)	36	Nemacur 3	68	Very highly toxic
Blue gill sunfish (*L. macrochirus*)	15	Nemacur 15G	151	Highly toxic
Rainbow trout (*Oncorhynchus* sp.)	15	Nemacur 15G	563	Highly toxic
Blue gill sunfish (*L. macrochirus*)	Not reported	Fenamiphos sulfone	1,173	Moderately toxic
Blue gill sunfish (*L. macrochirus*)	Not reported	Fenamiphos sulfoxide	2,653	Moderately toxic
Blue gill sunfish (*L. macrochirus*)	Not reported	Fenamiphos sulfoxide	2,000	Moderately toxic

Adapted from Patrick et al. (2001)

this area is required (Choo and Baker 1998). Megharaj et al. (2003a) reported that fenamiphos application to soil up to 10 kg/ha was not generally toxic to native soil algal flora or to soil enzyme activities involving dehydrogenase, phosphatase, and glucosidase. The toxicity of fenamiphos to several aquatic and terrestrial organisms is summarized in Table 8.

Table 8 Fenamiphos LC_{50} values for different aquatic and terrestrial species

Organisms	Reported dose	Source
Amphipod *Echinogammarus tibaldi*	11 μg/L	Pantani et al. (1997)
Scud *Gammarus italicus*	20 μg/L	Pantani et al. (1997)
Blue gill *Lepomis macrochirus*	0.0096 mg/L	Tomlim (2000)
Rainbow trout *Oncorhynchus mykiss*	0.0721 mg/L	Tomlim (2000)
Rainbow trout *Salmo gairdneri*	68 μg/L	Patrick et al. (2001)
Water flea *D. magna*	0.0019 mg/l	Tomlim (2000)
Alga *Scenedesmus subspicatus*	11 mg/L	Tomlim (2000)
Sheepshead minnow *Cyprinodon variegatus*	117 μg/L	Patrick et al (2000)
Worms *Eisenia foetida*	795 mg/kg soil	Tomlim (2000)

5.2.1 Effects on Wildlife

Fenamiphos and its major degradation products (FSO and FSO_2) are rated as "very highly toxic" to most terrestrial organisms. Limited quantities of these pesticides can impair reproductive capability or cause the death of wildlife. Terrestrial wildlife may be exposed to fenamiphos applied to the ground surface, through either deliberate or incidental ingestion of soil or granules, when feeding or preening. Wildlife may ingest residues directly from soil, ingest soil-living invertebrates and plants, or may inhale or have dermal contact with residues.

The level of concern (LOC) is a criterion for identifying the potential of a pesticide to cause adverse effects to nontarget organisms, even when properly applied according to label directions. After soil incorporation, it has been estimated that the concentration of fenamiphos exceeded the OP's LOC by several orders of magnitude. Wildlife deaths have been reported in the USA, after application of granular fenamiphos on grapes and golf courses. Incidences of bird deaths have occurred despite watering-in (irrigation) on turf to reduce surface availability of fenamiphos. One incident was recorded to have occurred when drip irrigation was performed at night to reduce attraction of birds to the irrigation water (Patrick et al. 2001). The toxicity of fenamiphos is very high to birds. The acute oral LD_{50} for the ring-necked pheasant is 0.5 mg/kg and varies from 1 to 2.4 mg/kg for other bird species. In a controlled experiment, fenamiphos appeared to be the most toxic agent among 13 different cholinesterase inhibitors. In tests with wild songbirds (red-winged blackbirds and house sparrows), an unspecified dose of Nemacur was highly toxic, with death of the birds occurring within 1 hr of eating the granules (Extoxnet 1996).

In 1990, the US EPA reported that the death of American robins (*Turdus migratorius*) and cedar waxwings (*Bombycilla cedrorum*) was linked to fenamiphos application to turf. Tissue sample analysis confirmed that the death of these birds resulted from the fenamiphos application. From 1994 to 1996, mitigation measures were implemented in the USA to reduce the risk related to fenamiphos use. However, in 1996, the EPA received a report about the death of 28 American coots (*Fulica americana*) from exposure to fenamiphos that had been applied to a golf course; the application followed label instructions. Based on these incidents, the EPA concluded that the use of Nemacur on turf can cause bird kills, even if the product is used in accordance with the then current label directions and restrictions (Patrick et al. 2001).

5.3 Toxicity and Fate in Mammals

5.3.1 Fate and Acute Toxicity

The effect of fenamiphos on different species of mammals has been studied under laboratory conditions. This pesticide, as other OP compounds do, inhibits the enzyme acetylcholinesterase. Its metabolites, fenamiphos sulfoxide and fenamiphos sulfone, showed a higher inhibition of the enzyme than did the parent compound (FAO/WHO 1994). These compounds can be absorbed through the skin, the gastrointestinal tract, or can be inhaled.

Exposure studies in rats using ^{14}C fenamiphos demonstrated that the compound is metabolized quickly to the sulfoxides and sulfones, *N*- and *O*-dealkylation products, and conjugates. Fenamiphos was excreted within 12–15 hr after administration of a single oral dose of 2 mg/kg bwt. The metabolites, excreted mainly through the urine, were identified as fenamiphos sulfoxide phenol sulfate, fenamiphos sulfoxide phenol, fenamiphos phenol sulfate and fenamiphos sulfone phenol, fenamiphos phenol, fenamiphos sulfoxide, 3-hydroxymethyl fenamiphos sulfone phenol sulfate, and desisopropyl fenamiphos sulfoxide. The metabolites found in the feces were fenamiphos sulfoxide, fenamiphos sulfoxide phenol, fenamiphos sulfone phenol, and fenamiphos phenol sulfate. Approximately 93% of the radioactive product applied was recovered and no parent compound was detected (FAO/WHO 1997). The acute toxicity of fenamiphos and its metabolites was determined in different animal species, mainly through research conducted at Bayer and toxicological evaluations submitted to the WHO. A summary of the acute mammalian toxicity for fenamiphos and its metabolites is presented in Table 9.

As part of an acute neurotoxicity screening test battery, male and female Wistar rats (18/sex/dose), fasted (overnight), were given a single real dose of fenamiphos at 0, 0.4, 1.6, and 4 g/kg. Twenty-five minutes and 1 d after administration,

Table 9 Acute toxicity of fenamiphos and its metabolites to different animals

Animal	Sex	Route	LD$_{50}$ mg/kg bwt
Fenamiphos Mouse	M	Oral	22.7
Rat	M	Intraperitoneal	3.4
	M	Oral	2.4
	M	Intraperitoneal	3
Guinea pig	M	Oral	56
Rabbit	M	Oral	10
	M	Intraperitoneal	17.3
Cat	M	Oral	10
Dog	M	Oral	10
Chicken	F	Oral	5.3
Fenamiphos sulfone Rat	M	Oral	2.6
Fenamiphos sulfoxide Rat	M	Oral	2.4
Desisopropyl fenamiphos Rat	M	Oral	1.4
Desisopropyl fenamiphos sulfone Rat	M	Oral	4.1
Fenamiphos sulfoxide phenol Rat	M	Oral	1,418
Fenamiphos sulfone phenol Rat	M	Oral	1,250
4-methylthio-meta-cresol Rat	M	Oral	1,418

bwt = body weight
Adapted from FAO/WHO (1997)

the rats were evaluated with behavioral testing; neuropathological examination was carried out at terminal sacrifice. Plasma, RBC (red blood cell), and brain cholinesterase activities were measured approximately 50 min post dosing in six rats/sex/dose. No treatment-related changes were observed in mean body weights, absolute and relative brain weights, and the incidence of gross and neurohistopathological lesions. At the higher dose, fenamiphos toxicity was observed within 21 and 31 min post dosing: (lethality 7/15 males and 1/12 females), with clinical signs of cholinesterase inhibition persisting for approximately 2 hr. Related effects, like muscle fasciculation, gait incoordination, nasal and oral staining, constricted pupils, salivation, and lacrimation, were also observed (Cruz 1999).

5.3.2 Long-Term Toxicity Studies

A study on long-term exposure to fenamiphos was conducted using Wistar rats. The rats were fed different dosages in the diet (30 mg/kg being the maximum administrated dose). The population was monitored for 2 mon. During the first 6 wk of the study, signs of cholinergic stimulation were observed. Inhibition of plasma cholinesterase was reported at a concentration of 3 mg/kg diet. Incremental increases in the weight of the thyroid gland were observed in females treated with the highest dose, but this tissue did not display any pathological abnormality. The NOEL was 3 mg/kg diet (equivalent to 0.17–0.23 mg/kg bwt/d) on the basis of plasma cholinesterase inhibition (FAO/WHO 1994). In a 2-yr chronic rat toxicity study in which animals were exposed to fenamiphos at a maximum concentration of 50 mg/kg diet, no carcinogenicity was observed.

Teratogenic studies were conducted using pregnant female rats. The animals were orally administered fenamiphos (92.5%) at doses of 0, 0.3, 1, and 3 mg/kg bwt/d from day 6 to day 15 of gestation. Cholinergic signs were observed in 18 out of 25 dams receiving 3 mg/kg bwt/d. The pesticide administered at a dose of 0.3 and 1.0 mg/kg bwt/d caused toxicity. It was concluded that fenamiphos, up to a concentration of 3 mg/kg bwt/d, was not embryotoxic or teratogenic in rats (FAO/WHO 1994). The effect of fenamiphos on the reproduction of rats was determined in a study wherein male and female rats were fed with moderate to high dosages of the compound (0.15–1.5 mg/kg/d). Over three generations, no compound-related reproductive effects were noted at the middle dose corresponding to 0.5 mg/kg/d. At a dose above 0.5 mg/kg/d, the second generation of pups showed a decrease in body weight gain. This effect was not seen in the third generation (Extoxnet 1996).

6 Isofenphos

Isofenphos (Fig. 3) is an OP insecticide used to control soil-dwelling insects such as white grubs, cabbage root flies, corn roundworms, and wireworms. The product is used in vegetables, including carrots, maize, on soil insects, in fruit crops like banana, and on turfgrass. On insects, it is a selective contact and stomach poison. It is applied as a preplant or preemergence soil treatment. Isofenphos is also used to control termites in and around structures. The insecticide may be found in formulations with the fungicide thiram. It is formulated as an emulsifiable concentrate, dry seed treatment, and granular and wettable powder (Extonet 1996). Large amounts of this insecticide was used in the USA until the year 2000, when it was withdrawn from the market by Bayer, the producer (US EPA 1999). However, the insecticide is still used extensively in other countries such as China (Yi et al. 2006; Li et al. 2007). In 2007, the Pesticide Residues Committee (2007) in the UK reported that residues of isofenphos methyl were found in Spanish peppers from Spain, in a survey conducted at supermarkets. This report was also confirmed by the authorities

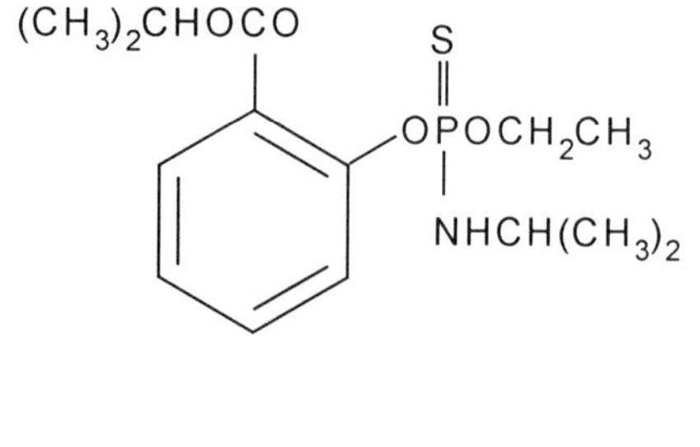

Fig. 3 Isofenphos

from Germany, Finland, and Holland. The Spanish horticultural authorities discovered from this event that approximately 60% of the isofenphos used by Spanish growers was imported from China, rather than bought in the EU.

6.1 Environmental Fate

6.1.1 Breakdown in Soil

Isofenphos is moderately to highly persistent in the soil environment. At a normal field-use rate, a single application of isofenphos had a half-life of 30–300 d, with a typical time of 150 d (Wauchope et al. 1992). Isofenphos is poorly bound to soils, but has low water solubility (Wauchope et al. 1992). Isofenphos slowly leaches in all soils and the breakdown products leach more rapidly than does the parent compound. The first step (oxidation) in the breakdown of the compound requires the presence of oxygen. The products of this process are isopropyl salicylate and cyclic isofenphos. Degradation was not greatly influenced by the addition of organic matter to the soil (Somasundaram et al. 1987). The evaporation of breakdown products accounts for a substantial portion of the loss of these residues from the soil. The effect of three soil pHs, three soil temperatures, and three soil moistures on ^{14}C isofenphos degradation was investigated by Abou-Assaf and Coats (1987). It was observed that the aforementioned factors interact in the degradation of isofenphos. However, the degradation was faster at higher pH and humidity. The formation of isofenphos oxon was greatest at higher temperatures and under acidic and neutral conditions.

Chapman and Harris (1982) studied the persistence of isofenphos in a mineral and an organic soil, and residues of the insecticide in vegetables grown in treated soils. Isofenphos residues declined slowly, with 50% of the initial application remaining at 12 wk in both soil types. One year after an application of isofenphos to sand and muck, 15 and 45% of the original residues were recovered, respectively, declining to 4 and 14%, respectively after 2 yr. Residues of isofenphos in radish and carrot generally did not exceed 0.04 ppm, with the exception of carrot (0.25 ppm) grown in isofenphos-treated sand. Chapman et al. (1986) reported a rapid degradation of isofenphos, mediated by microorganisms, in a clay loam soil that received repeated applications of the insecticide. Racke and Coats (1988) also reported enhanced biodegradation of isofenphos in soils that had had a long history of pesticide use. The microbial role in the degradation of isofenphos was demonstrated by Ohshiro et al. (1997). They isolated a bacterium from turf-green soil and extracted enzymes from the cells cytosol that rapidly hydrolyzed isofenphos. The enzymes cleaved an aryl phosphoester bond in the insecticide and produced its hydrolysates: *O*-ethyl isopropylphosphoramidothioate and isopropyl salicylate.

6.1.2 Breakdown in Water

The maximum concentration expected from runoff, following a field application of the insecticide isofenphos, was very low for water (0.07 mg/L) and sediment (0.04 mg/L) (US EPA 1990). The rate of isofenphos breakdown in the presence of water (hydrolysis) is markedly increased by acidic or basic conditions and by increases in temperature. Breakdown in the presence of sunlight is very slow. However, Zamy et al. (2004) reported that phototransformation of isofenphos in aqueous solutions led to the formation of four degradation products.

6.1.3 Breakdown in Vegetation

Plants, such as corn and onion, absorb isofenphos from the soil and, in corn, it translocates to leaves and stems. Plants generally transform the compound to its oxygen analog. The oxygen analog is then translocated acropetally within the plant, where further changes occur. Cereal grains, leafy and root vegetables, and edible oil crops grown in soil treated 9 mon earlier have been shown to take up the residues of insecticide even when it was present at very low concentrations (0.005 mg/kg). The principal residues detected in plants were isofenphos and an isofenphos-oxygen analog (US Public Health Service 1995).

6.2 Ecotoxicology

6.2.1 Effects on Aquatic Organisms

Isofenphos is moderately toxic to fish. The LC_{50} values of the compound for various species of fish are relatively consistent. The reported 96-hr LD_{50} values are 2 mg/L in goldfish, 2–4 mg/L in carp, 1–2 mg/L in orfe, and 1 mg/L in rudd (Kidd and James 1991). Study results indicate possible adverse effects to aquatic protozoa at exposure concentrations exceeding 20 mg/L (US EPA 1990). The potential of the compound to significantly bioaccumulate is relatively low. When channel catfish were exposed to 0.01 mg/L for a month, residues accumulated to 75 times the water concentration within the first 7 d, and thereafter declined during the treatment period. At the end of treatment cycle, fish tissue residues decreased by 87% within the first day and by 96% within 10 d. The extractable tissue residue was exclusively isofenphos (US EPA 1990).

6.2.2 Effects on Birds

Isofenphos is highly toxic to birds. The reported 5-d dietary LC_{50} in Japanese quail is 299 ppm (Smith 1993). In northern bobwhite quail, the reported acute LD_{50} is 13 and 19 mg/kg for technical grade and formulation products, respectively (Hill and Camardese 1984). For hens, the LD_{50} ranged between 6 and 20 mg/kg. Repeated dosages at 2 mg/kg/d for 90 d did not affect the nerves of hens. Red-winged blackbirds are also susceptible to isofenphos (Balcomb et al. 1984). In a laboratory study, the death of over 100 blackbirds was linked to application of Oftanol, a formulation of isofenphos.

6.3 Toxicity in Mammals

6.3.1 Acute Toxicity

Isofenphos is highly toxic with a reported acute oral LD_{50} of 28–38 mg/kg in rats and 91.3–127 mg/kg in mice (Kidd and James 1991; US Public Health Service Hazardous Substance Data Bank 1995). The product is moderately toxic via the dermal route, with a reported dermal LD_{50} of 188 mg/kg in rats (US National Institute for Occupational Safety and Health. Registry of Toxic Effects of Chemical Substances 1986; Kidd and James1991). The acute dermal LD_{50} in rabbits ranges from 162 to 315 mg/kg (US Public Health Service Hazardous Substance Data Bank 1995; Kidd and James 1991). The compound causes no damage to skin or mucous membranes in rats from topical applications; however, isofenphos is highly toxic via inhalation, with two reported airborne LC_{50s} of 1.3 and 0.144 mg/L, in rats (US Public Health Service Hazardous Substance Data Bank 1995; Kidd and James 1991). The hamster 4-hr inhalation LC_{50} is 0.23 mg/L (US Public Health Service Hazardous Substance Data Bank 1995). Typical of other organophosphorus insecticides, this compound is a cholinesterase inhibitor. Effects produced from exposure include increased secretions, difficulty in breathing, diarrhea, urination, pupil contraction, and slowing of the heartbeat (Broadberg 1990). At very high doses, convulsions and coma may ensue. Isofenphos may be more toxic when it is combined with the insecticide malathion (Broadberg 1990).

6.3.2 Long-Term Toxicity

The primary chronic effect of isofenphos in animals and in humans is suppression of cholinesterase activity. Dosages below 0.05 mg/kg/d in the diets of rats and mice had no effect on blood stream (plasma) cholinesterase activity. Rats and dogs fed 1.0 mg/kg/d of isofenphos for 3 mon exhibited no compound-related effects. Cholinesterase activity was not inhibited in rats fed a low dose of isofenphos (below 1.0 mg/kg/d) for 2 yr (Kidd and James 1991). Other tests on rats, however, showed evidence of nerve damage after repeated low-level exposures to some organophosphates (Wilson et al. 1984).

In 1990, an outbreak of ataxia occurred in over 700 pigs in the north of England. Follow-up studies demonstrated that the disorder was associated with the consumption of feed from a particular supplier and that one component (wheat) was common to the batch of feed with which the ataxia was associated. An analysis of the feed demonstrated the presence of an organophosphorus pesticide, later identified as isofenphos, an insecticide not approved for use in the UK. The wheat batches had been imported from France, and the warehouse in which they had been stored was contaminated with isofenphos, which is approved for restricted use in France. Isofenphos is known to cause delayed neuropathy. The dose to which the pigs were theoretically exposed would be expected to have resulted in neuropathy (manifested as ataxia) (Shaw et al. 1995).

6.3.3 Reproductive Effects

Female rats fed with isofenphos at dosages between 0.05 and 5.0 mg/kg/d through three successive litters exhibited no adverse effects on reproduction at the lowest dose, but maternal body weight reductions occurred at doses above 0.05 mg/kg/d. At 0.5 mg/kg/d, isofenphos produced decreases in reproductive success in some studies but not in others (US Public Health Service Hazardous Substance Data Bank 1995). However, some evidence of abnormalities associated with toxicity to the developing embryos was reported at 0.15 mg/kg/d (US Public Health Service Hazardous Substance Data Bank 1995). The available data on reproductive effects from exposure to isofenphos are inconclusive, but suggest that such effects are unlikely to be present in humans, at the expected exposure levels.

6.3.4 Teratogenic Effects

Pregnant rats fed with low to moderate amounts (0.3–3 mg/kg/d) of isofenphos during gestation had no compound-induced malformations in their offspring (US Public Health Service Hazardous Substance Data Bank 1995). Pregnant rabbits fed with isofenphos over a range of dosages 1–5 mg/kg/d during gestation had offspring with no teratogenic effects (US Public Health Service Hazardous Substance Data Bank 1995). These data indicate that isofenphos is not teratogenic.

6.3.5 Mutagenic Effects

The tests conducted on the mutagenic potential of isofenphos were all negative (Broadberg 1990; US Public Health Service Hazardous Substance Data Bank 1995). Additional tests using concentrations as high as 3.15 mL/plate confirmed the non-mutagenicity of isofenphos (Broadberg 1990; US Public Health Service Hazardous Substance Data Bank 1995).

6.3.6 Carcinogenic Effects

Mice fed on a diet of isofenphos at a dose of up to 5 mg/kg/d for 2 yr had no dose-related increases in tumors. However, there was a high mortality, not only in mice fed on isofenphos but also in the controls (Broadberg 1990; US Public Health Service Hazardous Substance Data Bank 1995). Rats fed on low to high doses of isofenphos (0.05–5 mg/kg/d) for 2 yr had no treatment-related alterations in 30 different tissues examined (Broadberg 1990). The evidence suggests that isofenphos is not carcinogenic. However, chronic organophosphate poisoning is associated with the development of agnogenic myeloid metaplasia and a rapid progression into acute myeloid leukemia. OP pesticides exert one of their carcinogenic effects through altering intermediary glucose metabolism, with myelomonocytic leukemia (CMML) having been documented to occur from exposure to isofenphos. In 2001, Boros and Williams reported that this pesticide exerts a leukemogenic effect by recruiting glucose carbons for nucleic acid synthesis, thus promoting cellular proliferation simultaneous with poor differentiation. The imbalanced metabolic phenotype demonstrates a severe defect in glucose oxidation and lipid and amino acid synthesis, concurrent with de novo synthesis of nucleic acids in response to isofenphos treatment. This profile conforms to the invasive proliferating phenotype observed in TGF-β-treated lung epithelial carcinoma cells. Further studies showed that the insecticide isofenphos exerts its leukemogenic effects in lymphocytes isolated from both males and females. It was found that overwhelming DNA mutation rates resulted from exposure to remarkably low concentrations of isofenphos and its metabolites, and these mutation levels exceeded endogenous repair capacity that resulted in formation of karyotypic instability (Williams et al. 2004). In summary, the main target organs affected by isofenphos exposure include the central and peripheral nervous systems and the blood.

6.4 Fate in Humans and Animals

The metabolism of isofenphos was examined in adults and larvae of southern corn rootworm (SCR), *Diabrotica undecimpunctata howardi* Barber SCR. Adults and larvae were analyzed at specific time intervals after treatment with ^{14}C isofenphos at the LD_{10} dosage level. Most of the ^{14}C was recovered from the extract of the internal organ and excreta fractions. The major metabolic pathways of isofenphos in SCR included oxidative desulfuration of isofenphos to isofenphos oxon and hydrolysis of isofenphos oxon and/or parent isofenphos to isopropyl salicylate. Isofenphos penetrated approximately three times faster into larvae than into adults. A lower accumulation of toxic compounds inside the larvae, as a result of a faster metabolism (1.5×) and more rapid elimination of the parent isofenphos, may explain why larvae are less susceptible than adults (Hsin and Coats 1986).

Rats, pigs, and cows all eliminate isofenphos rapidly. Rats excreted nearly all of the compound in urine and the remaining small amount in feces within 3 d of the initial dosing. Pigs eliminated 80% of the parent compound in urine and most of the remaining amount in feces within 1 d. Cows had a very similar pattern of excretion over a 2-d interval. Less than 1% of the initial dose was detected in milk (US Public Health Service Hazardous Substance Data Bank 1995). When domestic hens were fed a moderate dosage of isofenphos (4 mg/kg/d) daily for 3 d, a significant amount of the compound was eliminated within 2 d. Residues of the compound in the tissues and eggs were less than 3% of the administered dose. The major residues found in excreta, tissues, and eggs were isofenphos and isopropyl salicylate (US Public Health Service Hazardous Substance Data Bank 1995). Though the compound can be temporarily stored in tissues, the concentrations drop after exposure ceases. Rats fed on isofenphos at a high dose (15 mg/kg/d) for 6 d had small amounts of the compound in muscle, liver tissue, fat, and in the kidneys. Within 5 d postexposure, the concentration in these tissues had fallen to very low levels (Hudson et al. 1984). Cows had peak blood plasma levels of isofenphos 2 hr after dosing. Levels fell sharply within 1 d. The rate of metabolism of ^{14}C isofenphos (IFP) to isofenphos oxon (IFP-oxon), des N-isofenphos (d-N-IFP), and des N-isofenphos oxon (d-N-IFP-oxon) by rat, guinea pig, monkey, dog, and human liver microsomal p450 enzymes was studied (US Public Health Service 1995) to obtain V_{max} and K_m values for Michaelis–Menten kinetics. The monkey had the highest V_{max} value (162 nmol isofenphos per hour per 1.3 nmol p-450) for the conversion of IFP to IFP-oxon (desulfuration), followed by guinea pig (98 nmol), rat (66 nmol), dog (43 nmol), and human (14 nmol). The K_m values for the desulfuration of isofenphos were 19.2, 7.4, 14.1, 23.3, and 18.4 µM, respectively, for the monkey, guinea pig, rat, dog, and human. The V_{max} values (nmol isofenphos per hour per 1.3 nmol P-450) for the dealkylation process (conversion of IFP to d-N-IFP) were 64.6, 17.2, 9.7, and 7.3 for the monkey, rat, dog, and human, respectively. For the dealkylation process, monkey had the highest K_m value, 16.3 µM IFP, followed by human (11.2 µM), rat (9.9 µM), and dog (9.3 µM). The rate of metabolism of IFP-oxon and d-N-IFP to d-N-IFP-oxon was studied separately. The V_{max} and K_m values obtained in this study for animal and human liver p-450 enzymes are being used to develop a PB-PK/PB-PD (physiologically based pharmacokinetics/physiologically based pharmacodynamics) model to predict the fate and toxicity of isofenphos in animals and man (US National Library of Medicine 2004).

7 Coumaphos

Coumaphos (Fig. 4) is an organophosphorus insecticide used for control of a wide variety of livestock arthropods, including cattle grubs, screw worms, lice, scabies, flies, and ticks. It is used against ectoparasites, which are arthropods that live on the outside of host animals such as sheep, goats, horses, pigs, and poultry. It is added to cattle and poultry feed to control the development of fly larvae that breed in manure. It is also used as a dust, dip, or spray to control mange, horn flies, and face flies of cattle. Because of its low toxicity to fish, it is also used in water as an agent to control mosquito larvae. Coumaphos is considered a selective insecticide because it kills specific insect species while sparing other nontarget organisms (Extoxnet 1996).

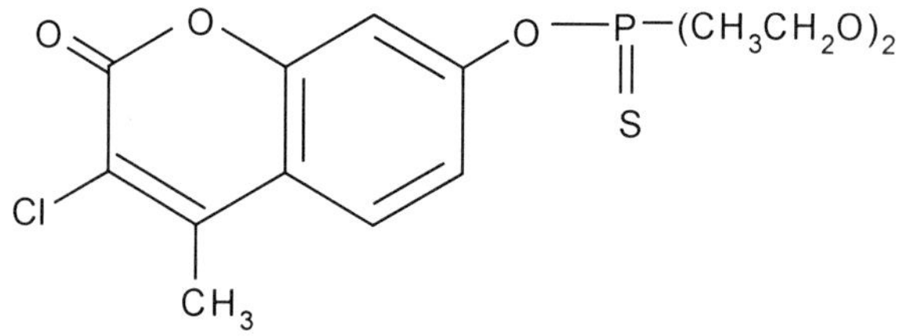

Fig. 4 Coumaphos

7.1 Environmental Fate

7.1.1 Breakdown in Soil

Based on the general characteristics of OP compounds, coumaphos is expected to have low to moderate persistence in soil. Coumaphos was relatively immobile in a sandy loam soil and is unlikely to contaminate groundwater. A general characteristic of OP compounds such as coumaphos is that they bind fairly well to soil particles. Therefore, they do not readily move (leach) with water percolating through the soil (US Public Health Service Hazardous Substance Data Bank 1995).

Jindal et al. (2000) studied the dissipation of coumaphos in an alkaline sandy loam soil in India for 1 yr. The study measured the degradation and leaching of ^{14}C coumaphos, alkylated ^{14}C coumaphos, and vat-aged residues of ^{14}C coumaphos in field soil columns under subtropical conditions. It was found that the dissipation, degradation, and bound residues formation in soil columns were more pronounced in alkali-treated coumaphos than in untreated coumaphos. After 365 d, 33.25% of ^{14}C-coumaphos equivalents were recovered as total residue from untreated coumaphos samples as compared to 19.12% from samples of alkali-treated coumaphos. Bound residue of coumaphos accumulated almost two times more in alkali treatments (18.95%) than in the other treatments (9.53%) at 150 d, but declined thereafter in both cases. The proportion of metabolites 4-methylumbelliferone, chlorferon, and potasan collectively was 86.05% of the extractable residue from untreated coumaphos, as compared to 91.74% from the alkali-treated coumaphos, after 365 d. Aged residues from vats containing copper sulfate and buffer (total residues recovered at 150 d – 95.6%) were found to be more persistent in the soil, than were the aged residues from vats containing only buffer (83.09% recovered). Copper sulfate seems to inhibit the degradation of coumaphos in soil by microorganisms. Chlorferon was the major metabolite of coumaphos in all the samples. Coumaphos did not leach below 10 cm in any of the mentioned studies.

Coumaphos can undergo degradation mediated by microorganisms. Horne et al. (2002) isolated a *Pseudomonas monteilii* strain (designated C11) capable of hydrolyzing coumaphos. The bacterium was isolated from soil of a cotton farm near Narrabri, NSW, Australia. Ha et al. (2007) isolated two consortia of bacterial cultures responsible for the degradation of chlorferon and diethylthiophosphate (DETP), the two main metabolites of coumaphos. These consortia of bacteria were isolated from the waste of a cattle dip solution. Mulbry (2000) also reported that *Nocardioides simplex* NRRL B-24074, a member of a coumaphos-degrading microbial consortium from cattle dip waste, was able to degrade coumaphos and other organophosphates. Shelton and Somich (1988) isolated coumaphos-degrading bacteria from cattle dips along the US–Mexican border. Three different types of bacteria were isolated and one of them was able to mineralize the pesticide, while the others transformed the compound to chlorferon and diethylthiophosphoric acid.

7.1.2 Breakdown in Water

Coumaphos is resistant to breakdown in water (hydrolysis). It is nearly insoluble in water and is stable over a wide pH range (US Public Health Service Hazardous Substance Data Bank 1995). Mallet and Volpe (1978) studied the stability of coumaphos in distilled water under different pH conditions. They observed that, in distilled water, the insecticide was more stable at neutral pH than at acidic or basic conditions, with a half-life of 347 d at pH 7.0, 33 d at pH 4.0, and 29 d at pH 8.5. Corta et al. (2000) found that coumaphos was rapidly degraded at pH 10, but was stable under neutral and acidic conditions. The degradation of coumaphos in aqueous media is influenced by temperature and UV radiation. Jindal et al. (2007) observed that volatilization, mineralization, and degradation of coumaphos in water suspensions increased with an increase in temperature and exposure to solar radiation, particularly when the UV component of the solar radiation existed. Major loss of the pesticide occurred through volatilization. The optimum temperature for the degradation of coumaphos was found to be 37°C.

7.2 Ecotoxicology

7.2.1 Effects on Aquatic Organisms

Coumaphos is moderately toxic to fish and highly toxic to aquatic invertebrates (US Public Health Service Hazardous Substance Data Bank 1995). The LC_{50} (96 hr) for coumaphos is 0.8 mg/L in channel catfish, 1.1 mg/L in largemouth bass, and 0.8 mg/L in walleye (Pimentel 1971; Kidd and James 1991). The LC_{50} (96 hr) for coumaphos is 5.9 mg/L in rainbow trout, 5 mg/L in bluegill sunfish, and 0.00015 mg/L in freshwater invertebrates (amphipods) (US EPA 1989a). Coumaphos tends to accumulate slightly in fish. For example, bluegill sunfish bioconcentrated coumaphos 331 times the ambient water concentration; however, mortality was high among the fish at the concentration tested (0.1 mg/L).

7.2.2 Effects on Birds

Coumaphos is highly toxic to birds (US EPA 1989b). The symptoms of acute toxicity in mallards given a dietary concentration of 29.8 mg/kg include spraddle-legged walking, wing twitching, wing drop, tearing of the eyes, and spread of the wings. These symptoms persisted in some survivors for up to 13 d and were accompanied by weight loss. Death usually occurred between 2 and 12 hr after treatment. Severe acute toxicity, and eventual death, was caused in hens after they were given daily oral dosages of 10 mg/kg/d, for 1–8 d. Hens given a single oral dose of 50 mg/kg recovered from the initial effects of cholinesterase inhibition and developed signs of delayed nerve poisoning (US Public Health Service 1995). The oral LD_{50} for coumaphos is 3 mg/kg in wild birds, 29.4 mg/kg in mallard ducks, 7.94 mg/kg in pheasants, and 14 mg/kg in chickens (Kidd and James 1991; US Public Health Service Hazardous Substance Data Bank 1995).

7.2.3 Effects on Other Organisms

Coumaphos poses a moderate hazard to honeybees and a slight hazard to other beneficial insects. Haarmann et al. (2002) studied the effect of coumaphos on honey bee, *Apis mellifera* L., focusing mainly on queen viability and health. The queen exposed to coumaphos weighed significantly less and had lower ovary weights than the control group queen. The highest coumaphos concentrations were observed in the queen cells and wax of the high-dose groups.

8 Future Research

The soil and water pollution resulting from the application of agrochemicals such as OP pesticides is of great concern worldwide. Efforts have therefore been made to investigate the interactions between these pollutants and nontarget organisms. In this direction, the information available on the environmental fate of fenamiphos and related OP compounds is considerable. Although a wealth of information exists on the isolation and characterization of the organophosphate hydrolysis gene (*opd*) of the bacterium involved in methyl parathion and parathion hydrolysis, virtually no comparable information is available on the genetics of fenamiphos hydrolysis. To our surprise, there are only two reports available on the isolation of fenamiphos-hydrolyzing pure bacteria, although limited information exists on the isolation of mixed bacterial cultures. Further research is also required to determine the reasons for the inability of fenamiphos-hydrolyzing bacteria to hydrolyze related pesticides, such as methyl parathion and parathion, and vice versa. This information can help to construct hybrid bacteria with broad specificity to degrade several related OP pesticides that could then be used in the bioremediation of OP compounds in the environment. Pesticide impact assessment tools, ranging from simple indices to complex simulation models, could be valuable to predict the environmental fate of a pesticide. However, these models require actual data that reflect the behavior of the agrochemicals under real conditions. That is the reason why studies that address the environmental fate of key pesticides, such as fenamiphos, in different environmental compartments are greatly needed. Overall, the available data on toxicity of certain OP pesticides are far from complete and often inconsistent. Therefore, assessment of risk for these chemicals requires additional studies that include innovative molecular ecological approaches. Also, precise toxicity evaluation of a wide taxonomic range of nontarget organisms is warranted.

9 Summary

In this review, we emphasize recent research on the fate, transport, and metabolism of three selected organophosphorus pesticides (fenamiphos, isofenphos, and coumaphos) in soil and water environments. This review is also concerned with the side effects of these pesticides on nontarget organisms. Despite the fact that fenamiphos is not very mobile, its oxides have been detected in the groundwaters of Western Australia. Most organophosphorus pesticides generally are chemically unstable and readily undergo microbial degradation in soil and water environments. Enhanced biodegradation of many organophosphorus pesticides upon their repeated applications to soil and water is well established. Myriads of soil microorganisms, bacteria in particular, exhibit an exceptional capacity to transform many organophosphorus pesticides. Fenamiphos can undergo rapid microbially mediated degradation via oxidation to its oxides (sulfoxide and sulfone) and eventually to CO_2 and water in soils, or via hydrolysis, in cultures of the soil bacterium, *Brevibacterium* sp. There is evidence for enhanced biodegradation of (i) isofenphos in soils with a long history of use and (ii) coumaphos in cattle dips by bacterial cultures to chlorferon and diethylthiophosphoric acid.

In this review, we also address the status of the current understanding of the acute and chronic toxicity, and reproductive, teratogenic, mutagenic, and carcinogenic effects of our targeted OP pesticides. We also address their fate in humans and animals. There have been widespread OP-poisoning episodes that of wildlife in Europe as a result of the constant use of OP compounds. To us this indicates that these pesticides should be used in a judicious manner to avoid such nontarget intoxications. As a result of our literature review on these OP compounds, we recommend additional future research on them utilizing up-to-date molecular methodology. Such research, we believe, will enhance the ability to predict where wildlife intoxications are most probable.

Acknowledgments This project was funded by the Australian Government through an IPRS scholarship in collaboration with UniSA and CRC CARE.

References

Abdollahi M, Ranjbar A, Shadnia S, Nickfar S, Rezai A (2004) Pesticides and oxidative stress: a review. Med Sci Monit 106:141–147.
Abou-Assaf N, Coats JR (1987) Degradation of [^{14}C] isofenphos in soil in the laboratory under different soil pHs, temperatures, and moistures. J Environ Sci Health B 22:285–301.
Ahmad R, Kookana RS, Alston AM (2001) Sorption of ametryn and imazethapyr in twenty-five soils from Pakistan and Australia. J Environ Sci Health B 36:143–160.
Ahmad R, Kookana R (2002) Role of the chemistry of soil organic carbon in pesticide sorption in soils. In: Proceedings of the 17th World Soils Congress, Thailand.
Alexander M (1981) Biodegradation of chemicals of environmental concern. Science 211:132–138.
Alexander M (2000) Aging, bioavailability and overestimation of risk from environmental pollutants. Environ Sci Technol 34:4259–4265.
Anderson PE, Nevremann K, Haidt H (1998) Accelerated Microbial Degradation of Nematicides in soil: Problem and its management. Bayer AG report.
Anderson BS, Phillips BM, Hunt JW, Worcester K, Adams M, Kapellas N, Tjeerdema RS (2006) Evidence of pesticide impacts in Santa Maria River watershed, California, USA. Environ Toxicol Chem 24:1160–1170.
Appleyard SJ (1995) Investigation of ground water contamination by fenamiphos and atrazine in a residential area: source and distribution of contamination. Ground Water Monit Rev 17:110–113.

Aston LS, Seiber JN (1997) Fate of summertime airborne organophosphate pesticide residues in the Sierra Nevada Mountains. J Environ Qual 26:1483–1492.

Australian Academy of Technological Sciences and Engineering (2002) Pesticide use in Australia. Australian Academy of Technological Sciences and Engineering.

Balcomb R, Stevens R, Bowen C (1984) Toxicity of 16 granular insecticides to wild caught songbirds. Bull Environ Contam Toxicol 33:302–307.

Banerjee BD, Seth V, Bhattacharya A, Pasha ST Chakraborty, AK (1999) Biochemical effects of some pesticides on lipid peroxidation and free-radical scavengers. Toxicol Lett 107:33–47.

Barik S and Sethunathan N (1978) Biological hydrolysis of parathion in natural ecosystems. J Environ Qual 7:346–348.

Barriuso E, Koskinen WC, Sadowsky MJ (2004) Solvent extraction characterization and bioavailability of atrazine residues in soil. J Agric Food Chem 52:6552–6556.

Baskaran S, Bolan NS, Rahman A, Tillman RW (1996) Pesticide sorption by allophanic and non-allophanic soils of New Zealand. NZ J Agric Res 39: 297–310.

Beynon KI, Wright AN (1969) Breakdown of the insecticide, Gardona, on plants and in soils. J Sci Food Agric 20:250–257.

Blain PG (2001) Adverse health effects after low level exposure to organophosphates. Occup Environ Med 58:689–690.

Bondarenko S, Gan J, Haver DL, Kabashima JN (2004) Persistence of selected organophosphate and carbamate insecticides in waters from a coastal watershed. Environ Toxicol Chem 23: 2649–2654.

Booth LH, O'Halloran K (2001) A comparison of biomarker responses in the earthworm *Aporrectodea caligiosa* to the OP insecticides diazinon and chlorpyrifos. Environ Toxicol Chem 20:2494–2502.

Boros LG, Williams RD (2001) Isofenphos induced metabolic changes in K562 myeloid blast cells. Leukemia Res 25:883–890.

Boxall AB, Jhonson P, Smith EJ, Sinclair CJ, Stutt E, Levi LS (2006) Uptake of veterinary medicines from soils into plants. J Agric Food Chem 54:2288–2297.

Broadberg, RK (1990) Estimation of Exposure of Persons in California to Pesticide Products Containing Isofenphos. California Department of food and agriculture. Division of Pest Management, Sacramento, CA, 5–75.

Buchanan D, Pilkington A, Sewell C, Tannahill SN, Kidd MW, Cherrie B, Hurley JF (2001) Estimation of cumulative exposure to organophosphate sheep dips in a study of chronic neurological health effects among United Kingdom sheep dippers. Occup Environ Med 58:694–701.

Cáceres T (1995) Degradación y disipación del acaricida chlorfenvinphos en baños de inmersión modelo. Tesis. Pontificia Universidad Católica del Ecuador, p.96.

Cáceres T, Ying GG, Kookana R (2002) Sorption of pesticides in banana production on soils of Ecuador. Aus J Soil Res 40:1085–1094.

Cáceres T, Megharaj M, Naidu R (2007) Toxicity of fenamiphos and its metabolites to the Cladoceran *Daphnia carinata*: the influence of microbial degradation in natural waters. Chemosphere 66:1264–1269.

Cáceres T, Megharaj M, Naidu R (2008a) Degradation of fenamiphos in soils collected from different geographical regions: the influence of soil properties and climatic conditions. J Environ Sci Health B 43:314–322.

Cáceres T, Megharaj M, Naidu R (2008b) Sorption of fenamiphos to different soils: the influence of soil properties. J Environ Sci Health B 43:605–610.

Cáceres T, Megharaj M, Naidu R (2008c) Toxicity and transformation of fenamiphos and its metabolites by two micro algae *Pseudokirchneriella subcapitata* and *Chlorococcum* sp. Sci Total Environ 398:53–59.

Cáceres T, Megharaj M, Naidu R (2008d) Biodegradation of the pesticide fenamiphos by ten different species of green algae and cyanobacteria. Curr Microbiol 57: 643–646.

Cáceres T, Megharaj M, Malik S, Beer M, Naidu R (2009). Hydrolysis of fenamiphos and its toxic oxidation products by *Microbacterium* sp. in pure culture and groundwater. Bioresource Technol 100:2732–2736.

Capowiez Y, Rault M, Mazzia C, Belzunces L (2004) Earthworm behavior as a biomarker – a case study using imidacloprid. Pedobiologia 47:542–547.

Casida J, Quistad G (2004) Organophosphate Toxicology: safety aspects of nonacetylcholinesterase secondary targets. Chem Res Toxicol 17:983–998.

Chambers J, Levi P (1992) Organophosphates: Chemistry, fate, and effects. Academic Press Inc. San Diego, CA.

Chapman RA, Harris CR (1982) Persistence of isofenphos and isazophos in a mineral and an organic soil. J Environ Sci Health B 17:355–361.

Chapman RA, Harris CR, Moy P, Henning K (1986) Biodegradation of pesticides in soil: Rapid degradation of isofenphos in a clay loam after a previous treatment. J Environ Sci Health B 21:269–276.

Choo LPD, Baker GH (1998) Influence of four commonly used pesticides on the survival, growth, and reproduction of the earthworm *Aporrectodea trapezoides* (Lumbricidae). Aus J Agric Res 49:1297–1303.

Christensen OM, Mather JG (2003) Pesticide-induced surface migration by lumbricid earthworms in grassland: life-stage and species differences. Ecotoxicol Environ Saf 57:89–99.

Chu X, Wu NF, Deng MJ, Tian J, Yao B, Fan YL (2006) Expression of organophosphorus hydrolase OPHC2 in *Pichia pastoris*: purification and characterization. Prot Expr Purificat 49:9–14.

Chung KY, Ou LT (1996) Degradation of fenamiphos sulfoxide and fenamiphos sulfone in soil with a history of continuous applications of fenamiphos. Arch Environ Contam Toxicol 30:52–458.

Cooke CM, Shaw G, Lester JN, Collins CD (2004) Determination of solid–liquid partition coefficients (K_d) for diazinon, propetamphos and *cis*-permethrin: implications for sheep dip disposal. Sci Total Environ 329:197–213.

Corta E, Bakkali A, Barranco A, Berruela AL, Gallo B, Vicente F, Bogdanov S (2000) Study of the degradation products of bromopropylate, chlordimeform, coumaphos, cymiazole, flumethrin and tau-fluvalinate in aqueous media. Talanta 52:169–180.

Cruz J (1999) Human health risk assessment: fenamiphos. US Environmental Protection Agency. Office of Pesticide Programs. Health Effects Division (www.epa.gov/oppsrrd1/op/fenamiphos/hedsep99.pdf).

Dantoine T, Debord J, Merle L, Charmes, JP (2003) Roles of Paraoxonase 1 in organophosphate compounds toxicity and in atherosclerosis. Rev Med Int 24:436–442.

DebMandal M, Mandal S, Pal NK, Aich A (2008) Potential metabolites of dimethoate by bacterial degradation. World J Microbiol Biotechnol 24:69–72.

EXTOXNET (1996) Pesticide Information Profiles- Fenamiphos. (www.extoxnet.orst.edu/cgiwebglimpse/mfs/services/data/info/extoxnet?link=http://ace).

FAO/WHO (1997) Toxicological and Environmental Evaluations- Fenamiphos. Joint FAO/WHO, Lyon. (www.inchem.org/documents/jmr/jmpmno/v097pr06.htm).

FAO/WHO (1994) WHO/FAO Data sheet on pesticides N92, fenamiphos Joint FAO/WHO, Lyon. (www.inchem.org/documents/pds/pds/pest92_e.tm).

Farahat TM, Abdelrasoul GM, Amr MM, Shebl MM, Farahat FM, Anger WK (2003) Neurobehavioural effects among workers occupationally exposed to organophosphorous pesticides. Ocup Environ Med 60:279–286.

Felsot AS (1984) Persistance of isofenphos (Amaze) soil insecticide under laboratory and field conditions and tentative identification of a stable oxygen analog metabolite by gas chromatography. J Environ Sci Health B 19:13–27.

Feigenbrugel V, Le Person A, Le Calve S, Mellouki A, Muñoz A, Wirtz A (2006) Atmospheric fate of dichlorvos: photolysis and OH-initiated oxidation studies. Environ Sci Technol 40:850–857.

Feng Y, Park JH, Voice TC, Boyd SA (2000) Bioavailability of soil-sorbed biphenyl to bacteria. Environ Sci Tech 34:1977–1984.

Franzmann P, Zappia L, Tilbury A, Patterson B, Davis G, Mandelbaun T (2000) Bioaugmentation of Atrazine and Fenamiphos impacted groundwater: laboratory evaluation. Bioremediat J 43:237–248.

Gan J, Papiernik SK, Koskinen WC, Yates R (1999) Evaluation of accelerated solvent extraction (ASE) for analysis of pesticide residues in soil. Environ Sci Technol 33:3249–3253.

Gao JP, Maguhn J, Spitzauer P, Kettrup A (1998) Sorption of pesticides in sediment of the Teufelsweiher pond (Southern Germany) I: equilibrium assessment, effect of organic carbon content and pH. Water Res 32:1662–1672.

Guerin WF, Boyd SA (1992) Differential bioavailability of soil-sorbed naphthalene to two bacterial species. Appl Environ Microbiol 58:1142–1152.

Guitart R, Sachana M, Caloni F, Croubels S, Vandenbroucke V (2009) Animal poisoning in Europe. Part 3: Wildlife. Vet J (in press). doi:10.1016/j.tvjl.2009.03.033.

Gultekin F (2000) The effect of organophosphate insecticide chlorpyrifos-ethyl on lipid peroxidation and antioxidant enzymes (in vitro). Arch Toxicol 74:533–538.

Gupta RC (2001) Depletion of energy metabolites following acetylcholinesterase inhibitor-induced status epilepticus: protection by antioxidants. Neurotoxicol 22:271–282.

Ha J, Engler CR, Wild JR (2007) Biodegradation of chlorferon and diethylthiophosphate by consortia enriched from waste cattle dip solution. Biores Technol 98(10):1916–1923.

Haarmann T, Spivak M, Weaver D, Weaver B, Glenn T (2002) Effects of fluvalinate and coumaphos on queen honey bees (Hymenoptera: Apidae) in two commercial queen rearing operations. J Econ Entomol 95:28–35.

Harmsen J, Rulkes W, Eijsackers H (2005) Biovailability: concept for understanding or tool for predicting? Land Contam Reclam 13:161–171.

Hatzinger PB, Alexander M (1995) Effect of aging of chemicals in soil on their biodegradability and extractability. Env Sci Tech 29:537–545.

Hill EF, Camardese MB (1984) Toxicity of anticholinesterase insecticides to birds: Technical grade versus granular formulations. Ecotoxicol Environ Saf 8:551–563.

Horne I, Harcourt RL, Sutherland TD, Russell RJ, Oakeshott JG (2002) Isolation of a *Pseudomonas monteilii* strain with a novel phosphotriesterase. FEMS Microbiol Lett 206(1):51–55.

Hsin CY, Coats JR (1986) Metabolism of isofenphos in southern corn rootworm. Pestic Biochem Physiol 25(3):336–345.

Hudson RH, Tucker RK, Haegele MA (1984) Handbook of Toxicity of Pesticides to Wildlife. Resource Publication 153. US Department of Interior, Fish and Wildlife Service, Washington, DC, 5–16.

IAEA (International Atomic Energy Agency) (1991) Laboratory Training Manual on the Use of Nuclear and Associated Techniques in Pesticide Research. Division of Nuclear Techniques in Food and Agriculture. Technical Reports Series N 329, Vienna.

Jamal GA, Hansen S, Pikington A, Buchanan D, Gillham RA, Abdel- Aziz M, Julu POO, Al-Rawas F, Hurley F, Ballantyne JP (2002) A clinical neurological, neuropsychological study of sheep farmers and dippers exposed to organophosphate pesticides. Occup Environ Med 9:434–441.

Jiang J, Zhang R, Li R, Gu D, Li S (2007) Simultaneous biodegradation of methyl parathion and carbofuran by a genetically engineered microorganism constructed by mini-Tn5 transposon. Biodegradation 18:403–412.

Jindal T, Singh DK, Agarwal HC (2000) Persistence, degradation and leaching of coumaphos in soil. J Environ Sci Health B 35:309–320.

Jindal T, Singh DK, Agarwal HC (2007) Effect of UV radiation and temperature on mineralization and volatilization of coumaphos in water. J Environ Sci Health B 42:367–372.

Juhasz AL, Megharaj M, Naidu R (2000) Bioavailability: the major challenge (constrain) to bioremediation of organically contaminated soils. In: Remediation of Hazardous Waste Contaminated Soils 2nd ed. Vol. 1: Engineering Considerations and remediation Strategies, Section 1–1: Engineering Issues in Waste Remediation, pp 217–241.

Karpouzas D, Karanasios E, Menkissoosh U (2004a) Enhanced microbial degradation of cadusafos in soil from potato monoculture: demonstration and characterization. Chemosphere 56: 549–559.

Karpouzas D, Hatziapostolou P, Papadopoulou E, Giannakou I, Georgiadou A (2004b) The enhanced biodegradation of fenamiphos in soils from previously treated sites and the effect of soil fumigants. Environ Toxicol Chem 23:2099–2107.

Karpouzas D, Walker A (2000) Factors influencing the ability of *Pseudomonas putida* strains epI and II to degrade the organophosphate ethoprophos. J Appl Microbiol 89:40–48.

Karpouzas D, Morgan J, Walker A (2000) Isolation and characterization of ethoprophos-degrading bacteria. FEMS Microbiol Ecol 33:209–218.

Karpouzas DG, Fotopoulou A, Menkissoglu-Spiroudi U, Singh K (2005) Non-specific biodegradation of the organophosphorus pesticides, cadusafos and ethoprophos, by two bacterial isolates. FEMS Microbiol Ecol 53:369–378.

Katan J, Lichtenstein EP (1977) Mechanism of production of soil bound residues of ^{14}C parathion by microorganisms. J Agric Food Chem 25:1404–1408.

Kavalci C, Durukan P, Ozer M, Cevik Y, Kavalci G (2009) Organophosphate poisoning due to a wheat bagel. Inter Med 48:85–88.

Kearney PC, Karns JS, Muldoon MT, Ruth JM (1986) Coumaphos disposal by combined microbial and UV-ozonation reactions. J Agric Food Chem 34: 702–706.

Kelsey JW, Alexander M (1997) Declining bioavailability and inappropriate estimation of risk of persistent compounds. Environ Toxicol Chem 16:582–585.

Kelsey JW, Colino A, White JC (2005) Effect of species differences, pollutant concentration, and residence time in soil on the bioaccumulation of 2,2, bis(*p*-chlorophenyl)-1,1-diclorothlene by three earthworms species. Environ Toxicol Chem 24:703–708.

Kidd H, James DR (1991) The Agrochemicals Handbook. Royal Society of Chemistry Information Services, 3rd edn.Cambridge, UK, 5–14.

Kikuchi M, Sasaki Y, Wakabayasi M (2000) Screening of organophosphate insecticide pollution in water by using *Daphnia magna*. Ecotoxicol Environ Saf 47:239–245.

Koleli N, Demir A, Arslan H, Kantar C (2007) Sorption behavior of methamidophos in a heterogeneous alluvial soil profile. Colloids Surf A Physicochem Eng Asp 301:94–99.

Kookana R, Di H, Aylmore L (1995) A field study of leaching and degradation of nine pesticides in a sandy soil. Aus J Soil Res 33(6):1019–1030.

Kookana R, Phang C, Aylmore AG (1997) Transformation and degradation of fenamiphos nematicide and its metabolites in soils. Aus J Soil Res 35(4):753–762.

Koskinen WC, Harper J (1990) The retention process: mechanisms. In: Cheng HH (ed) Pesticides in the soil environment: process, impacts and modeling. 77, Soil Science Society of America, Madison, WI, pp 51–77.

Kraus M, Wilcke W, Zech W (2000) Availability of polycyclic hydrocarbons (PAHs) and poplychlorinated biphenyls (PCBs) to earthworms in urban soils. Environ Sci Technol 34: 4335–4340.

Landrum M, Cañas JE, Coimbatore G, Cobb GP, Jackson WA, Zhang B, Aderson TA (2006) Effects of perchlorate on earthworm (*Eisenia fetida*) survival and reproductive success. Sci Total Environ 363:237–244.

Lanno R, Wells J, Conder J, Bradman K, Basta N (2004) The bioavailability of chemicals in soil to earthworms. Ecotoxicol Environ Saf 57:39–47.

Lawrence MAM, Davies NA, Edwards PA, Taylor MG, Simkiss K (2000) Can adsorption isotherms predict sediment bioavailability? Chemosphere 41:1091–1100.

Levi PE, Hodgson E (1992) Metabolism of organophosphorus compounds by the flavin-containing monooxygenase. In: Chambers JE, Levi PE (eds) Organophosphates: chemistry, fate and effects, Academic Press Inc. San Diego, CA. pp 141–154.

Leyhe J (2004) Coumaphos, análisis of risk to endangered and threatened salmon and steal head. US EPA. http://www.epa.gov/espp/litstatus/effects/coumaphos/coumaphos_analysis.pdf.

Li L, Qian Ch, Jiang S, Zhou Z, Pan C (2007) Determination of organophosphorus pesticides in *Lycium barbarum* by gas chromatography with flame photometric detection. JAOAC 90: 271–276.

Liste HH, Alexander M (2002) Butanol extraction to predict bioavailability of PAHs in soil. Chemosphere 46:1011–1017.

Liu YH, Liu Y, Chen ZS, Lian J, Huang X, Chung YC (2004) Purification and characterization of a novel organophosphorus pesticide hydrolase from *Penicillium lilacinum* BP303. Enzyme Microbial Technol 34:297–303.

Long JLA, House WA, Parker A, Rae JE (1998) Micro-organic compounds associated with sediments in the Humber rivers. The Sci Total Environ 210:229–253.

Mallet VN, Volpe Y (1978) Degradation of coumaphos in distilled water as a function of pH. Anal Chimi Acta 97:415–418.

McCall PJ, Swann RL, Bauriedil WR (1985) Volatility characteristics of chlorpyrifos from soil. Rep. GH-C 1782. Dow Chemical USA, Midland, MI.

McConnell L, Nelson E, Rice C, Baker J, Jhonson E, Harman J, Bialek K (1997) Chlorpyrifos in the air and surface water of Chesapeake Bay: Predictions of atmosphecric depositions fluxes. Environ Sci Technol 31:1390–1398.

Meharg A (1996) Bioavailability of atrazine to soil microbes in the presence of the earthworm *Lumbricus terrestris* (L.). Soil Biol Biochem 28:555–559.

Megharaj M, Singh N, Kookana S, Naidu R, Sethunathan N (2003a) Hydrolysis of fenamiphos and its oxidation products by a soil bacterium in pure culture, soil and water. Appl Microbiol Biotechnol 61:252–256.

Megharaj M, Singleton I, Kookana S, Naidu R (2003b) Persistence and effects of fenamiphos on native algal populations and enzymatic activities in soil. Soil Biol Biochem 31:1549–1553.

Morrison DE, Robertson BK, Alexander M (2000) Bioavailability to earthworms of aged DDT, DDE, DDD and dieldrin in soil. Environ Sci Technol 34:709–713.

Mosleh Y, Ismail SM, Ahmed M, Ahmed Y (2003) Comparative toxicity and biochemical responses of certain pesticides on the mature earthworm *Aporrectodea caliginosa* under laboratory conditions. Environ Toxicol: 18:338–346.

Mulbry WW, Karns J, Keraney PC, Nelson JO, McDaniel CS, Wild JR (1986) Identification of a plasmid-borne parathion hydrolase gene from *Flavobacterium* sp. by Southern hybridization with *opd* from *Pseudomona diminuta*. Appl Environ Microbiol 51:926–930.

Mulbry WW (2000) Characterization of a novel organophosphorus hydrolase from *Nocardiodes simplex* NRRL B-24074. Microbiol Res 154:285–288.

National Pesticide Information Centre (1994) OSU Extension Pesticide Properties Database. http://npic.orst.edu/ppdmove.htm

Oliver D, Kokana R, Satama R (2003) Land use effects on sorption of pesticides and their metabolites in sandy soils. I. Fenamiphos and two metabolites, fenamiphos sulfoxide and fenamiphos sulfone, and fenarimol and azinphos methyl. Aus J Soil Res 41:847–860.

Oliver GR, McKellar RL, Woodburn KB, Eger JE, McGee GG, Ordiway TR (1987) Field dissipation and leaching study for chlorpyrifos in Florida citrus. Report No. Gh-C 1870. Dow Chemical USA, Midland, MI.

Ohshiro K, Ono T, Hoshino T, Uchiyama T (1997) Characterization of isofenphos hydrolases from *Arthrobacter* sp. strain B-5. J Ferment Bioeng 83:238–245.

Ortega S, Holliman PJ, Jones DL (2006) Toxicology and fate of Pestanal and commercial propetanphos formulations in river and estuarine sediments. Sci Total Environ 366:826–836.

Ou LT, Rao PSC (1986) Degradation and metabolism of oxamyl and phenamiphos in soils. J Environ Sci Health B 21:25–40.

Ou LT (1991) Interactions of microorganisms and soil during fenamiphos degradation. Soil Sci Soc Am J 55:716–722.

Ou LT, Thomas JE, Dickson DW (1993) Enhanced biodegradation of the nematicide fenamiphos in soil. In: Sorption and Degradation of pesticides and organic chemicals in soil. Soil Sciences Society of America, spec publ. 32. SSSA and ASA, Madison, WI, pp 253–260.

Ou LT, Thomas JE, Dickson DW (1994) Degradation of fenamiphos in soil with a history of continuos fenamiphos applications. Soil Sci Soc Am J 58:1139–1147.

Pakala SB, Gorla P, Pinjari AB, Krovidi RK, Baru R, Yanamandra M, Merrick M, Siddavattam D (2007) Biodegradation of methyl parathion and *p*-nitrophenol: evidence for the presence of a *p*-nitrophenol 2-hydroxylase in a Gram negative *Serratia* sp. strain DS001. Appl Microbiol Biotechnol 73:1452–1462.

Pantani C, Pannuncio G, Cristofaro M, Novelli A, Salvatori M (1997) Comparative acute toxicity of some pesticides, metals, and surfactants to *Gammarus italicus* Goedm. and *Echinogammarus tibaldii* Pink and Stock (Crustacea: Amphipoda). Bull Environ Contam Toxicol 59:963–967.

Patrick G, Chiri A, Randall D, Libelo L, Jones J (2001) Fenamiphos Environmental Risk Assessment. US Environmental Protection Agency. (http://www.epa.gov/oppsrrd1/op/fenamiphos/envrisk.pdf).

Pazy Miño C, Bustamante G, Sanchez ME, Leone P (2002) Cytogenetic monitoring in a population occupationally exposed to pesticides in Ecuador. Environ Health Persp 11:1077–1080.

Pesticide Residues Committee (2007) Isofenphos-methyl in Spanish peppers – melon survey replaced by peppers survey. Programme for pesticide residues in the UK food supply. http://www.pesticides.gov.uk/prc.asp?id=2150.

Phillips BM, Anderson BS, Hunt JW, Huntley SA, Tjeerdema RS, Kapellas N, Worcester K (2006) Solid phase sediment toxicity identification evaluation in an agricultural stream. Environ Toxicol Chem 25:1671–1676.

Pimentel D (1971) Ecological Effects of Pesticides on Nontarget Species. Executive Office of the President's Office of Science and Technology. US Government Printing Office, Washington, DC, 5–28.

Racke KD (1992) Degradation of organophosphorus insecticides in environmental matrices. In: Chambers JE, Levi PE (eds) Organophosphates: chemistry, fate and effect. Academic Press Inc., San Diego, CA, pp 47–73.

Racke KD, Coats JD (1987) Enhanced degradation of isofenphos by soil microorganisms. J Agric Food Chem 35:193–199.

Racke KD, Coats JR (1988) Comparative degradation of organophosphorus insecticides in soil-specificity of enhanced microbial degradation. J Agric Food Chem 36:93–199.

Racke KD, Coats JR (1990) Enhanced biodegradation of pesticide in the environment. ACS Symposium Series 426. American Chemical Society. Washington, DC.

Ragnarsdottir K (2000) Environmental fate and toxicology of organophosphate pesticides. J Geol Soc London 157:859–876.

Ranjbar A, Pasalar P, Abdollahi M (2002) Induction of oxidative stress and acetylcholinesterase inhibition in organophosphorus pesticide manufacturing workers. Hum Exp Toxicol 21: 179–182.

Rao JV, Pavan SY, Madhavendra SS (2003) Toxic effects of chlorpyrifos on morphology and acetylcholinesterase activity in the earthworm, *Eisenia foetida*. Ecotoxicol Environ Saf 54:296–301.

Raushel F (2002) Bacterial detoxification of organophosphate nerve agents. Curr Opin Microbiol 5:288–295.

Reginato JB, Koskinen WC, Sadowsky M (2006) Influence of soil aging on sorption and bioavailability of simazine. J Agric Food Chem 54:1373–1379.

Reid BJ, Jones KC, Semple KT (2000) Bioavailability of persistant organic pollutants in soils and sediments- a perspective on mechanisms, consequences and assessment. Environ Pollut 108:103–112.

Ritter WF, Jhonson HP, Lovely WG, Olnau M (1974) Atrazine, propachlor and diazinon residues on small agricultural watersheds: runoff losses, persistence and movement. Environ Sci Technol 8:38–42.

Schmidt-Nielsen K (1983) Animal Physiology: Adaptation and Environment. 3rd edn. Cambridge University Press, Cambridge.

Semple KT, Doick KJ, Jones KC, Burauel P, Craven A, Harms H (2004) Defining bioavailability and bioaccessability of contaminated soil and sediment is complicated. Environ Sci Tech 38:228–231.
Sethunathan N, Pathak MD (1972) Increased biological hydrolysis of diazinon after repeated application in rice paddies. J Agric Food Chem 20:586–589.
Sethunathan N (1973) Degradation of parathion in flooded acid soils. J Agric Food Chem 21: 602–604.
Sethunathan N, Yoshida T (1973) A *Flavobacterium* sp. that degrades diazinon and parathion as sole carbon source. Can J Microbiol 19: 873–875.
Sethunathan N, Siddaramappa R, Rajaram KP, Barik S, Wahid PA (1977) Parathion: residues in soil and water. Residue Rev 68: 91–122.
Shaw IC, Parker RM, Porter S, Quick MP, Lamont MH, Patel RK, Norman IM, Johnson MK (1995) Delayed neuropathy in pigs induced by isofenphos. Vet Res 136:95–97.
Shelton DR, Karns JS (1988) Coumaphos degradation in cattle-dipping vats. J Agric Food Chem 36:831–834.
Shelton DR, Somich CJ (1988) Isolation and characterization of coumaphos-metabolizing bacteria from cattle dip. Appl Environ Microbiol 54: 2566–2571.
Simon L, Spiteller M, Haisch A, Wallnofer P (1992) Influence of soil properties on the degradation of the nematocide fenamiphos. Soil Biol Biochem 24:769–773.
Singh N, Megharaj M, Gates WP, Churchman GJ, Anderson J, Kookana RS, Naidu R, Chen Z, Slade P, Sethunathan N (2003a) Bioavailability of an organophosphorus pesticide, fenamiphos, sorbed on an organoclay. J Agric Food Chem 51:2653–2658.
Singh B, Walker A, Morgan J, Wright D (2003b) Role of soil pH in the Development of enhanced biodegradation of fenamiphos. Appl Environ Microbiol 69:7035–7043.
Singh B, Walker A, Wright D (2005) Cross-enhancement of accelerated biodegradation of organophosphorus compounds in soils: Dependence on structural similarity of compounds. Soil Biol Biogeochem 37:1675–1682.
Singh B, Walker A (2006) Microbial degradation of organophosphorus compounds. FEMS Microbiol Rev 30:428–471.
Smith GJ (1993) Toxicology and pesticide use in relation to wildlife: Organophosphorus and carbamate compounds. CK Smoley, Boca Raton, FL, pp. 5–7.
Somasundaram L, Racke KD, Coats JR (1987) Effects of Manuring on the Persistence and Degradation of Soil Insecticides. Bull Environ Contam Toxicol 39:579–586.
Spencer WF, Shoup TD, Cliath MM, Farmer WJ, Haque R (1979) Vapor pressures and relative volatility of ethyl and methyl parathion. J Agric Food Chem 27:273–278.
Stenersen J, Brekke E, Engelstad F (1992) Earthworms for toxicity testing; species differences in response towards cholinesterase inhibiting insecticides. Soil Biol Biochem 24: 1761–1764.
Stone DL, Sudakin DL, Jenkins JJ (2009) Longitudinal trends in organophosphate incidents reported to the National Pesticide Information Center, 1995–2007. Environ Health 8:18.doi:10.1186/1476-069X-8-18.
Sultatos L (1992) Role of glutathione in the mammalian detoxification of organophosphorus insecticides. In: Chambers JE, Levi PE (eds) Organophosphates: chemistry, fate and effect. Academic Press Inc., San Diego, CA, pp 155–168.
Thiele-Bruhn S, Brummer GW (2004) Fractionated extraction of polycyclic aromatic hydrocarbons (PAHs) from polluted soils: estimation of PAH fraction degradable through bioremediation. Eur J Soil Sci 55:567–578.
Tomlin CDS (2000) The pesticide manual: a world compendium. Farnham: British Crop Protection Council, UK.
US Environmental Protection Agency (1989a) Pesticide Fact Sheet Number 207: Coumaphos. Office of Pesticides and Toxic Substances, Washington, DC, pp 5–19.
US Environmental Protection Agency (1989b) Registration Standard for Pesticide Products Containing Coumaphos as the Active Ingredient. Office of Pesticide Programs, Washington, DC, pp 5–55.

US Environmental Protection Agency (1990) Pesticide Environmental Fate One-Line Summary: Isofenphos. Environmental Fate and Effects Division, Washington, DC, pp 5–77.

US EPA (1999) Isofenphos; Receipt of Request to Cancel Registrations (http://www.epa.gov/EPA-PEST/1999/January/Day-15/p1024.htm).

US National Institute for Occupational Safety and Health. 1981–1986 Registry of toxic effects of chemical substances. Cincinnati, OH. US National Library of Medicine. Tox NET. (http://toxnet.nlm.nih.gov).

US National Library of Medicine (2004) Specialized Information Services. SIS, US http://toxnet.nlm.nih.gov/cgi-bin/sis/.

US Public Health Service (1995) Hazardous Substance Data Bank. Washington, DC, pp 5–9.

Valverde García A, Socías Viciana AM, González Pradas E, Villafranca Sánchez M (1992) Adsorption of chlorpyrifos on Almería soils. Sci Total Environ 123:541–549.

Van der Wal L, Tjallig J, Fleuren R, Barendregt A, Sinnige, Cornelis A, Van Gestel M, Hermens J (2004) Solid-phase microextraction to predict bioavailability and accumulation of organic micropollutants in terrestrial organisms after exposure to a field-contaminated soil. Environ Sci Technol 38:4842–4844.

Vermeulen LA, Reinecke AJ, Reinecke SA (2001) Evaluation of the fungicide manganese-zinc ethylene bis (dithiocarbamate) (Mancozeb) for sublethal and acute toxicity to *Eisenia fetida* (Oligochaeta). Ecotoxicol Environ Saf 48:183–189.

Verrell P, Buskirk E (2004) As the worm turns: *Eisenia fetida* avoids soil contaminated by a glyphosate-based herbicide. Bull Eviron Contam Toxicol 72:219–224.

Waggoner TB (1972) Metabolism of Nemacur [ethyl 4(methylthio)-m-tolyl isopropylphosphoramidate] and identification of two metabolites in plants. J Agric Food Chem 20:157–160.

Watschke EG, Mumma RO (1989) The effect of nutrients and pesticides applied to turf on the quality of runoff and percolating water. Environ. Resources Res. Inst. Report ER 8904. Pennsylvania State University, USA.

Wauchope RD (1978) The pesticide content of surface water draining from agricultural fields – a review. J Environ Qual 7:459–472.

Wauchope, RD, Buttler TM, Hornsby AG, Augustijn-Beckers PW, MBurt JP (1992) SCS/ARS/CES Pesticide properties database for environmental decision making. Rev Environ Contam Toxicol 123:1–157.

Wauchope RD, Yeh S, Linders JB, Kloskowski R, Tanaka K, Rubin B, Katayama A, Kordel W, Gerstl Z, Lane M, Unsworth JB (2002) Pesticide soil sorption parameters: theory, measurement, uses, limitations and reliability. Pest Manag Sci 58:419–445.

Weber J, Wilkerson G, Reinhardt C (2004) Calculating pesticide sorption (Kd) using selected pesticide sorption coefficients (Kd) using selected soil properties. Chemosphere 55:157–166.

WHO (1974) Fenamiphos. WHO Pesticide Residues series 4. (www.inchem.org/documents/jmr/jmpmno/v074pr23.htm).

Williams RD, Boros LG, Kolanko CJ, Jackma SM, Eggers TM (2004) Chromosomal aberrations in human lymphocytes exposed to the anticholinesterase pesticide isofenphos with mechanisms of leukemogenesis. Leukemia Res 28:947–958.

Wilson BW, Hooper M, Chow E, Higgins RJ, Knaack JB (1984) Antidotes and neuropathic potential of isofenphos. Bull Environ Contam Toxicol 33:386–394.

Xu G, Li Y, Zheng W, Peng X, Li W, Yan Y (2007) Mineralization of chlorpyrifos by co-culture of *Serratia* and *Trichosporon* spp. Biotechnol Lett 29:1469–1473.

Yang C, Dong M, Yuan Y, Huang Y, Guo X, Qiao C (2007) Reductive transformation of parathion and methyl parathion by *Bacillus* sp. Biotechnol Lett 29:487–493.

Yi X, Hua Q, Lu Y (2006) Determination of organophosphorous pesticides residues in the roots of *Platycclom grandiflorum* by solid phase extraction and gas chromatography with flame photometric detection. J AOAC Internat 89:225–231.

Zamy CC, Mazellier P, Legube B (2004) Phototransformation of selected organophosphorus pesticides in dilute aqueous solutions. Water Res 38:2305–2314.

Zhang Z, Hong Q, Xu J, Zhang X, Li S (2006) Isolation of fenitrothion-degrading strain *Burkholderia* sp. FDS-1 and cloning of *mpd* gene. Biodegradation 17:275–283.

Index

A

Abiotic parameters, gammarid toxicity
  effects, 18
Acetylcholinesterase
  activity, neurotoxicity biomarker, 50
  inhibition, OP mode-of-action, 119
Acute poisoning, OP (organophosphorous)
  pesticides, 132
Acute toxicity, fenamiphos in fish, (table), 140
Animals, fenamiphos acute toxicity
  (table), 143
Aquatic ecotoxicology, *Gammarus* spp., 1 ff
Aquatic species effects
  coumaphos, 152
  isofenphos, 146
Aquatic species toxicity, fenamiphos
  (table), 141
Arctic avian species, use as bioindicators, 105
Avian species, bioindicators, 105
Avian species effects
  coumaphos, 152
  isofenphos, 146

B

Behavioral assays, gammarids, 22
Behavioral test methods, gammarid toxicity
  (table), 23
Behavior testing, drift and foraging activity, 29
Bioaccumulation in gulls
  feeding affects, 87
  influential factors, 86
Bioavailability of OP pesticides
  earthworms, 130
  microbes, 130
Bioavailability from soil, OP pesticides, 129
Bioenergetic responses, gammarids, 32
Bioindicator species
  Arctic birds, 105
  in the European Arctic, Svalbard glaucous
    gull, 77 ff
Biological degradation, OP pesticides, 127
Biomarker(s)
  acetylcholinesterase, 50
  for detoxification, glutathione-S-
    transferase, 52
  in gammarids, vitellogenin-like
    proteins, 45
  in *Gammarus* spp. (table), 38
  metallothioneins and lipid peroxidation, 49
  Svalbard glaucous gulls, 89
  tests, gammarids, 32
Biomonitoring system, multispecies freshwater
  biomonitor (MFB), 27
Birds
  effects of coumaphos, 152
  effects of isofenphos, 146
Breakdown in water, coumaphos, 151
Brominated flame retardants, Svalbard
  glaucous gulls, 81

C

Carbamate insecticides, neurotoxic
  biomarker, 50
Carcinogenicity, isofenphos, 148
Chemical classes, OP pesticides, 119
Chemical contamination, European Arctic, 78
Chemical properties, fenamiphos (table), 134
Chemical toxicity, freshwater invertebrates and
  fish (table), 9
Chemical toxicity, gammarids, 7
Chemistry, OP pesticides, 118
Chiral legacy organochlorines, glaucous
  gulls, 81
Chitin and molting, gammarid biomarker, 46
Chronic toxicity
  fenamiphos, 143
  isofenphos, 147

Coastal sediment toxicity, *G. locusta*, 8
Contaminant(s)
    effects, Svalbard glaucous gulls (table), 89
    exposure, thermoregulatory effects, 97
    genotoxicity, Svalbard glaucous gulls, 100
    levels and patterns, Svalbard glaucous gulls, 79
    vs. nest temperature, Svalbard glaucous gulls (illus.), 98
    research, Svalbard glaucous gulls, 77 ff
    Svalbard glaucous gull residues, 84
Contaminated sediment toxicity, *G. pulex*, 8
Contaminate effects, on avian species, 104
Contaminate list, definitions (table), 109
Coumaphos
    aquatic species effects, 152
    avian species effects, 152
    description, 150
    ecotoxicology, 152
    environmental fate, 150
    fate in water, 151
Culturing gammarids, methods, 6

D
Daphnia, post-exposure feeding depression, 20
DDT contamination, European Arctic, 80
Degradation
    of isofenphos, vegetation, 146
    in soil and water, fenamiphos, 134
Depuration, gammarid toxicity, 19
Drift behavior, *Gammarus*
    and pollution, 22
    and predators, 22

E
Earthworms, bioavailability of OP pesticides, 130
Ecological biomarkers, Svalbard glaucous gulls, 89
Ecotoxicology
    coumaphos, 152
    fenamiphos, 139
    Gammarus spp., 1 ff
    isofenphos, 146
Embryotoxicity in gammarids, pollutants, 42
Emerging test species, *Gammarus* spp., 64
Endocrine disruption in Gammarids, endpoints, 45
Environmental fate
    coumaphos, 150
    fenamiphos, 134
    OP pesticides, 117 ff., 121
European Arctic
    chemical contamination, 78
    organochlorine contamination, 80
    Svalbard glaucous gull bioindicator species, 77 ff
Excretion rate, gammarids, 32
Exposure modes, assessing gammarids (table), 54
Exposure types, gammarids, 53
Extraction methods, organic pollutants in soil, 131

F
Fate
    in animals, isofenphos, 149
    in humans, isofenphos, 149
    in mammals, fenamiphos, 142
    in soil, isofenphos, 154
    in water, coumaphos, 151
    in water, isofenphos, 145
Feeding activity
    testing, gammarids, 10
    methods (table), 11
Feeding rate and uptake, gammarid toxicity, 19
Fenamiphos
    acute toxicity
        animals (table), 143
        aquatic and terrestrial species (table), 141
    chronic toxicity, 143
    description, 133
    ecotoxicology, 139
    environmental fate, 134
    fate and toxicity, 117 ff
    mammalian fate and toxicity, 142
    and metabolites
        leaching behavior, 137
        soil sorption (tables), 136
    microbial effects, 137
    physiochemical properties (table), 134
    plant metabolism, 138
    soil behavior, 135
    water residues (table), 135
    wildlife effects, 141
Fish toxicity, fenamiphos and metabolites (table), 140
Food choice experiments, *Gammarus*, 16
Freshwater invertebrates and fish, chemical toxicity (table), 9

G
Gametogenesis activity, *Gammarus* spp., 48
Gammaridea, natural habitat, 2
Gammarid(s)
    behavioral assays, 22
    bioenergetic responses, 32

Index

biomarker(s)
  chitin and molting, 46
  heat shock proteins, 45, 48
effects, pulsed exposure models, 57
endocrine disruption endpoints, 45
excretion and respiration rates, 32
exposure
  assessment, modes (tables), 54
  types, 53
feeding activity
  testing, 10
  and rate, modeling, 19
  test methods (table), 11
  toxicity effects, 18
feeding ecology, parasite effects, 21
in situ testing approaches, 58
in situ vs. *ex situ* test results, 18
laboratory *vs.* field tests, 18
lethality testing, 7
metallothionein induction, effects, 42
mode-of-action
  studies, 32
  test methods (table), 33
parasites, antipredator effects, 26
predation effects, 20
reproduction, pesticide effects, 44
sediment toxicity assays, 57
sensitivity, metal toxicity, 8
test methods, evaluating existing ones, 62
time-response assays, testing method, 10
toxicity
  behavioral test methods (table), 23
  feeding rate, uptake and depuration, 19
  PAHs (polyaromatic hydrocarbons), 8
  population dynamics effect, 43
  testing, feeding activity (table), 11
use in biomarker tests, 32

Gammarus
  aquatic ecotoxicology, 1 ff
  behavior and character, 2
  behavior testing, 4
  biomarkers (table), 38
  culturing methods, 5
  drift behavior
    and pollution, 22
    and predators, 22
  emerging test species, 5
  feeding behavior, 3
  food choice experiments, 16
  gametogenesis activity, 48
  as indicator species, 5
  leaf-mass feeding assays, 17
  life cycle, 2
  mating behavior, 3
  multimetric test systems, 64
  species distribution, 2
  toxicity endpoints, 4
  in water quality testing, 1 ff
Genotoxicity of contaminants, Svalbard glaucous gulls, 100
Glaucous gulls
  legacy chiral organochlorines, 81
  organohalogen contaminates (illus.), 80
G. locusta, coastal sediment toxicity, 8
Glucocorticoid effects, Svalbard glaucous gull residues, 96
Glutathione-S-transferase activity, detoxification biomarker, 52
Gonadal steroid hormones, Svalbard glaucous gull effects, 95
G. pulex, contaminated sediment toxicity, 8
Groundwater residues, fenamiphos (table), 135

H

Half-lives in soil, OP pesticides (table), 130
Hazards to humans and wildlife, OP pesticides, 118
Health effects, OP pesticides, 132
Heat shock proteins, gammarid biomarkers, 45, 48
Hormone effects, Svalbard glaucous gull residues, 94
Hydrolysis
  affects, OP pesticides, 126
  half lives, OP pesticides (table), 124

I

Immunity, Svalbard glaucous gulls, 98
In situ testing approaches, gammarids, 58
In situ vs. *ex situ* test results, gammarids, 18
Isofenphos
  aquatic species effects, 146
  carcinogenicity, 148
  chronic toxicity, 147
  description, 144
  effects on birds, 146
  fate
    in humans and animals, 149
    in soil, 145
    in water, 145
  mammalian toxicity, 147
  mutagenicity, 148
  reproductive toxicity, 148
  teratogenic effects, 148

L

Laboratory *vs.* field tests, gammarids, 18
Leaching behavior, fenamiphos, 137

Leaching potential, OP pesticides, 125
  table, 124
Leaf-mass feeding assays, *Gammarus*, 17
Lethality testing, gammarids, 7
Life cycle, *Gammarus*, 2
Lipid peroxidation, biomarker for metal
  exposure, 49
Long-term toxicity, fenamiphos, 143

M

Mammalian fate and toxicity, fenamiphos, 142
Mammalian toxicity, isofenphos, 147
Mercury residues, Svalbard glaucous gulls, 86
Metabolic enzymes, vertebrate effects, 93
Metabolism
  OP pesticides, 121
  in plants, fenamiphos, 138
Metabolites, Svalbard glaucous gulls, 82
Metallothionein(s)
  induction in gammarids, effects, 42
  and metal exposure, biomarkers, 49
Metal toxicity
  gammarids, 7
    sensitivity, 8
Microbial bioavailability, OP pesticides, 129
Microbial degradation, OP pesticides, 127
Microbial effects, fenamiphos, 137
Mode-of-action
  OP pesticides, 119
  studies, gammarids, 32
  test methods, gammarids (table), 33
Modeling, gammarid feeding activity and
  rate, 19
Multimetric *Gammarus* spp. tests,
  perspectives, 64
Multiple stressor biomarkers, *Gammarus* spp.
  (table), 38
Multispecies freshwater biomonitor (MFB),
  biomonitoring system, 27
Mutagenicity, isofenphos, 148

N

Nemacur, fenamiphos, 133
Nematicide description, fenamiphos, 133
Nematode infestation *vs.* PCB levels, Svalbard
  glaucous gulls (illus.), 100
Neurotoxic biomarker, acetylcholinesterase, 50
Neurotoxic effects, OP pesticides, 132
Neurotoxic mechanism, OP pesticides, 119

O

Occupational poisoning, OP pesticides, 132
OP mode-of-action, acetylcholinesterase
  inhibition, 119

OP (organophosphorous) pesticides,
  description, 118
OP pesticides
  acute poisoning, 132
  affect of hydrolysis, 126
  biological degradation, 127
  chemistry, 118
  environmental fate and transport, 121
  factors affecting bioavailability, 129
  health effects, 132
  human and wildlife hazards, 118
  hydrolysis half lives (table), 124
  leaching potential, 125
    table, 124
  metabolism, 121
  microbial bioavailability, 129
  microbial degradation, 127
  mode-of-action, 119
  neurotoxic biomarker, 50
  neurotoxic effects, 132
  occupational poisoning, 132
  oxidation/reduction, 126
  photolytic stability, 126
  run-off potential, 125
  soil
    adsorption values (table), 124
    bioavailability, 129
    half-lives (table), 130
  sorption to soil, 122
  structural types, 119
    illustration, 120
  volatility, 124
Organic pollutants in soil, extraction
  methods, 131
Organochlorine(s)
  contamination, Svalbard glaucous gulls, 84
  in glaucous gulls, legacy chiral forms, 81
  gull residues over time (illus.), 85
  legacy contaminants, 79
  site-specific accumulation effects, 88
Organohalogen contaminants, male glaucous
  gulls (illus.), 80
Organometals, Svalbard glaucous gulls, 83
Organophosphorous (OP) pesticides, fate and
  toxicity, 117 ff
Organotins, Svalbard glaucous gulls, 84
Oxidation/reduction, OP pesticides, 126
Oxidative stress, metal exposure, 49
Oxychlordane effects, gull breeding affects
  (illus.), 104

P

PAHs (polyaromatic hydrocarbons), gammarid
  toxicity, 8

Parasite(s)
    effects, gammarid feeding ecology, 21
    Svalbard glaucous gulls, 98
PBBs (polybrominated biphenyls), Svalbard glaucous gulls, 81
PBDE (polybrominated dephenyl ether)
    residues, Svalbard glaucous gulls, 86
    Svalbard glaucous gulls, 81
    vertebrate metabolism, 82
PCBs (polychlorinated biphenyls)
    contaminants, European Arctic, 79
    levels vs. nematode infestation, Svalbard glaucous gulls (illus.), 100
    thyroxin ratio effects (illus.), 95
    vertebrate metabolism, 82
Persistence in soil, OP pesticides (table), 130
Pesticide(s)
    effects, gammarid reproduction, 44
    exposure types, gammarids, 53
    fate and toxicity of OPs, 117 ff
    toxicity, gammarids, 7
PFASs (per- and poly fluorinated alkyl substances), Svalbard glaucous gull residues, 83
PFOS (perfluorooctane sulfonate), Svalbard glaucous gull residues, 83
Photolytic stability, OP pesticides, 126
Physical properties, fenamiphos (table), 134
Plant
    degradation, isofenphos, 146
    metabolism, fenamiphos, 138
Pleopod beat frequency, pollutant bioassay, 29
Pollutant
    bioassay, pleopod beat frequency, 29
    effects in gammarids, embryotoxicity, 42
Population dynamics effect, gammarids, 43
Porphyrins, Svalbard glaucous gulls, 93
Post-exposure feeding depression, *Daphnia*, 20
Predation effects, gammarids, 20
Prolactin effects, Svalbard glaucous gull residues, 96
Pulsed exposure models, gammarid effects, 57

R
Reproduction effects, Svalbard glaucous gull residues, 101
Reproductive behavior effects, Svalbard glaucous gull residues, 102
Reproductive toxicity, isofenphos, 148
Residue accumulation affects, Svalbard glaucous gulls, 88
Respiration rate, gammarids, 32

Retinoid effects, Svalbard glaucous gull metabolism, 93
Runoff potential, OP pesticides, 125

S
Sediment toxicity assays, gammarids, 57
Soil
    adsorption values, OP pesticides (table), 124
    behavior, fenamiphos, 135
    bioavailability, OP pesticides, 129
    degradation
        coumaphos, 150
        fenamiphos, 134
    fate, isofenphos, 145
    persistence, OP pesticides (table), 130
    sorption
        fenamiphos and metabolites (tables), 136
        OP pesticides, 122
        processes, OP pesticides, 123
Structural types, OP pesticides (illus.), 120
Surfactant toxicity, gammarids, 7
Svalbard glaucous gull
    bioaccumulation and age, 87
    bioindicator species, 77 ff
    breeding effects, oxychlordane residues (illus.), 104
    brominated flame retardants, 81
    chemical-induced effects, 78
    contaminant(s)
        effects (table), 89
        genotoxicity, 100
        levels and patterns, 79
        vs. next temperature (illus.), 98
    contamination, 84
    ecological biomarkers, 89
    effects, immunity and parasites, 98
    gender and bioaccumulation, 86
    hydroxylated metabolites, 82
    mercury residues, 86
    metabolic enzyme effects, 93
    metabolism, retinoids, 93
    organochlorine residues over time (illus.), 85
    organometals, 83
    PBBs, 81
    PBDEs, 81
        residues, 86
    PFAS residues, 83
    PFOS residues, 83
    porphyrins, 93
    residues

Svalbard glaucous (*cont.*)
    accumulation affects, 88
    gonadal steroid hormones, 95
    hormone effects, 94
    reproduction effects, 101
    reproductive behavior effects, 102
    thermoregulatory effects, 97
  trace elements, 83

T

Teratogenic effects, isofenphos, 148
Terrestrial species toxicity, fenamiphos (table), 141
Test methods for gammarids, evaluating existing ones, 62
Threshold level effects, contaminants and avian species, 104
Thyroxin ratio effect, PCBs (illus.), 95
Time-response assay methods, gammarids, 10
Toxicant effects, gammarid exposure modes (table), 54
Toxic effects, behavioral changes in MFB, 27
Toxicity
  in animals, fenamiphos (table), 143
  of chemicals, gammarids, 7
  effects
    on feeding, gammarid test methods (table), 11
  gammarid feeding activity, 18
  in fish, fenamiphos and metabolites (table), 140
  to gammarids, behavioral test methods (table), 23
  in mammals, fenamiphos, 142
  OP pesticides, 117 ff
Trace elements, Svalbard glaucous gulls, 83
Transport
  in environment, OP pesticides, 121
  processes, OP pesticides, 122

V

Vegetation, isofenphos breakdown, 146
Vertebrate(s)
  metabolic enzyme effects, 93
  metabolism, PCBs and PBDEs, 82
Vitellogenin-like proteins, gammarid biomarkers, 45
Volatility, OP pesticides, 124

W

Water
  -induced degradation, fenamiphos, 134
  quality testing, *Gammarus* spp., 1 ff
  residues, fenamiphos (table), 135
Wildlife effects, fenamiphos, 141